Y0-CQL-830

WIND ENERGY

WIND ENERGY

Tom Kovarik
Charles Pipher
John Hurst

Chicago • New York

Wind Energy
Copyright © 1979 Quality Books, Inc.
Published by Domus Books
400 Anthony Trail
Northbrook, IL 60062

3 4 5 6 7 8 9 10

Graphs, maps and illustrations by Carla A. Hurst—Carla Bach Art

Manufactured in the United States of America

Library of Congress Cataloging in Publication Data

Kovarik, Thomas J.
Wind energy.

Bibliography: p. 116
Includes index.
1. Wind power. 2. Electric power production.
I. Pipher, Charles, joint author. II. Hurst, John A., joint author. III. Title.
TK1541.K67 621.312'136 78-24721
ISBN 0-89196-034-1
ISBN 0-89196-032-5 pbk.

CONTENTS

PREFACE

For thousands of years the wind has occupied a special corner of the human mind. People have prayed to it, cursed it, written poems and ballads to it. And they have used it. Early sailors plied the oceans of the world, their ships driven by what wind their crude sails could catch. On land, farmers used simple windmills to pump water, irrigate and drain fields, and grind cereal grains. Over the centuries these simple machines improved in efficiency and sophistication. By the second half of the last century, small multi-bladed, water-pumping windmills sprouted like weeds on the great American prairie. The new century saw the widespread use of electricity and, in rural areas, reliance on electrical power generated by windmills.

As the present energy crisis has deepened, many leaders have taken a second look at windmills as an alternative source of power. And they have begun to see more than nostalgic rural settings or quaint pictures of the Zuider Zee. They see the answer to the world's energy needs. Many countries are providing extensive financial support for wind energy research and development. All over the world, private enterprises, independent research organizations, and universities are involved, too, improving existing technology and rapidly developing new methods of harnessing the power of the wind. Their work ranges almost as far as the wind itself, from developing more efficient, wind-driven generators for use in individual houses to creating giant machines that may some day power entire cities.

All of this money and effort is well worth it, for the wind is an awesome source of power within our grasp. According to one recent estimate, the power generated by wind-driven generators thinly scattered across 3 percent of the U.S. land area (that has the best wind patterns) would be equal to this country's *total* consumption of energy in 1972. And that power would be eternal, never-ending. Our constantly increasing energy needs and our shaky reliance on fossil fuels mandate the development of such renewable sources of power to ensure the energy needs of future generations both near and distant.

Wind Energy tells the tale of wind power—past, present, and future. It reviews the history of wind energy from the first simple machines to the giant wind-driven generators of today and gives a glimpse at the technology of the future. It defines the characteristics of wind and tells how wind prospecting, both by using scientific instruments and by observing changes in vegetation, is used to select good windmill sites. It shows how the blades of a windmill extract power from the wind and how performance and efficiency goals have led to the development of new and distinct types of machines. And, it shows how the power produced by these machines can be stored for future use.

Although *Wind Energy* is not specifically designed as a do-it-yourself manual, it includes many design guidelines, rules-of-thumb, ideas, and hints to help the innovative individual in choosing a wind energy system. The Appendix—Source/Resource—graphically describes available components for complete wind energy systems.

Authentic windmill built by Dutch family near Superior, WI
(original painting by Phil Austin, AWCS)

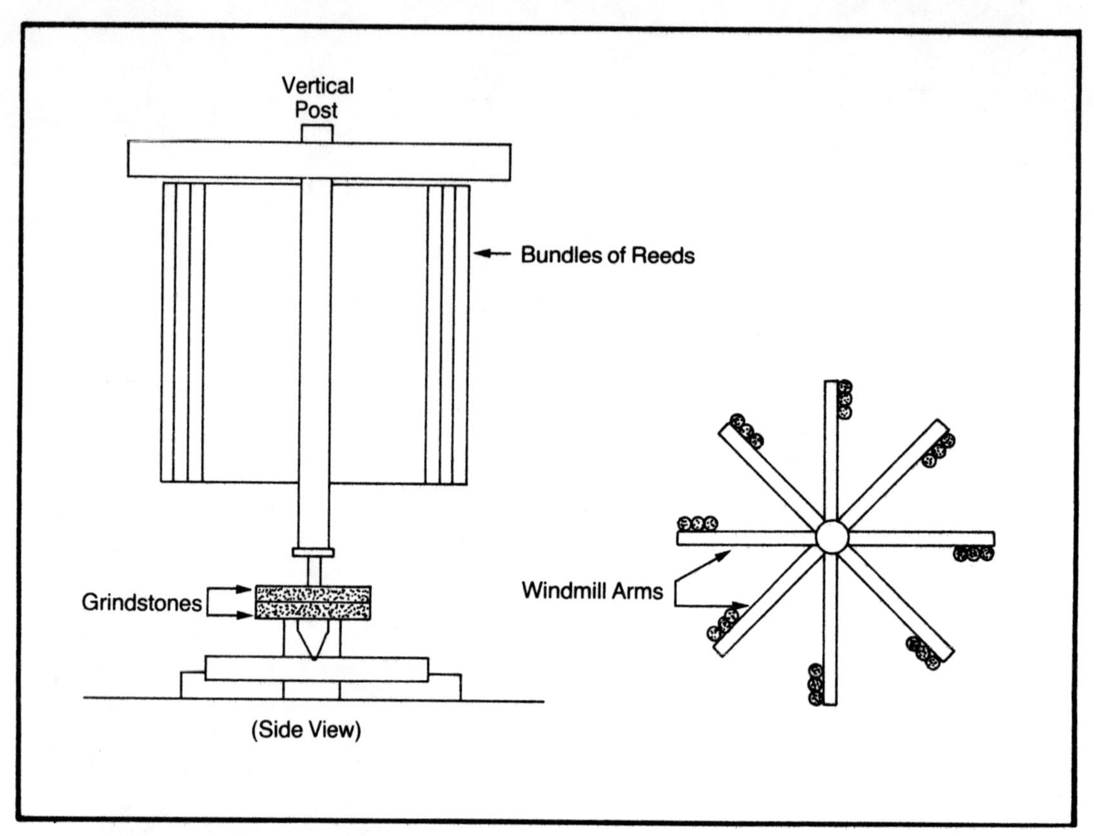

Fig. 1–1: How Persians built windmills in Seistan

Fig. 1–2: Two varieties of the simple but elegant post mill.

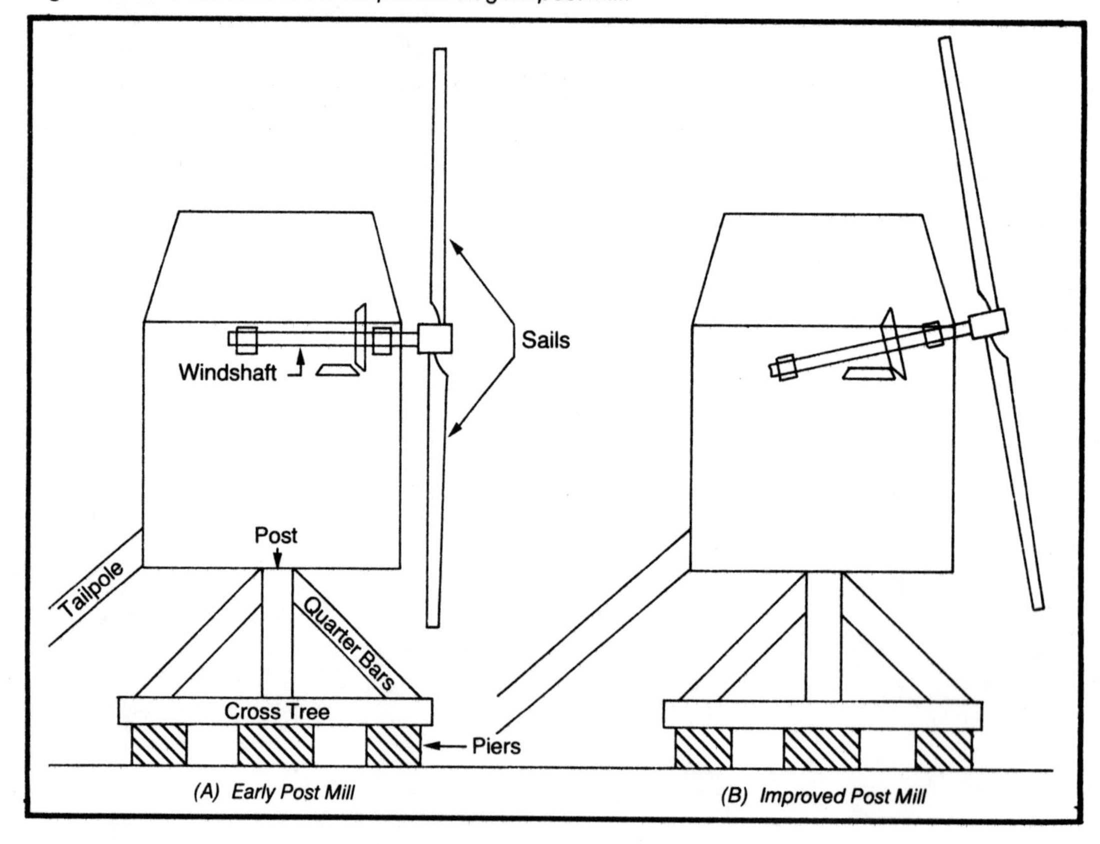

Chapter One

THE HISTORY OF WIND POWER

No one knows who invented the windmill, or when and where it first appeared, but any bets by scholars would put the Chinese, Babylonians, or the Persians out in front, and the date would be at least 2000 years ago, perhaps even double that. Most scholars accept this view since the first application of wind power that is recorded—that of the sail on small boats—can be traced back at least 5000 years to water craft on the Nile River.

The first windmills were vertical-axis machines with a number of arms on which sails were mounted, at first made from bundles of reeds, which were pushed around by the force of the wind (Fig. 1–1). They resembled a carousel or merry-go-round, and were used with millstones for grinding grain. It is not difficult to imagine why this form of windmill was the first to evolve. It is based on the timeless vertical post horizontal beam mill pushed or pulled around by men or animals walking in a circular path.

While some scholars believe that the Chinese were using simple vertical-axis windmills to pump irrigation water as long ago as 2000 B.C., other researchers give the Persians credit for the first large scale development of windmills, between 100 B.C. and 700 A.D. By the latter date windmills were a common sight in the Near East, and windmill building was a recognized and high-paying craft. Later, in the Mediterranean region, the first horizontal-axis windmills appeared, with wind wheels made of up to ten wooden booms and rigged with cloth jib sails, again a development probably linked to the sails of watercraft.

The first existing record of windmills in Europe dates to about 1100 A.D., and windmills were mentioned in English deeds late in the same century. By the fourteenth century windmills were a common sight throughout Europe and England, and many of these were post mills. The post mill was a remarkable machine, being changed little over a span of 600 years. Illustrations of post mills at work in 1275 by artists of that period differ little from modern photographs of the last few that were built during the 1800's and still survive today. All post mills (Fig. 1–2) have a shanty-like body, called the millhouse, supported by a vertical wooden post that in turn rests on two horizontal wooden beams. The wooden post is held upright by angled beams called quarterbars. The horizontal beams are called a *crosstree.* It rests on a stone foundation or piers to prevent the wood from rotting.

The millhouse contains the windshaft that connects to the sails and the mechanism that drives the grindstone. A long beam, called a *tailpole,* and often a ladder drop to the ground at the rear. To rotate the sails into the wind the entire millhouse is turned by moving the tailpole.

No one knows who invented the simple but elegant post mill, but some historians believe that the idea for the post mill was brought to Europe by returning crusaders who may have viewed simple horizontal-axis windmills near the Mediterranean. These mills were largely built of stone and fixed to face the prevailing winds along the coast. These early horizontal-axis windmills along the Mediterranean coast could not be rotated to face winds from other directions. The post mill was the first European development of a horizontal-axis windmill that could be turned to meet wind from any direction. Again, there is a strong tie to the timeless vertical post-horizontal beam mill pushed or pulled around by men or animals walking in a circular path.

Early post mills had the perfectly horizontal shafts of the first Mediterranean mills, but because of the stronger, variable and gusty winds of Europe they quickly caused excess wear on the wooden bearing block of the windshaft behind the sails (Fig. 1–3). As the sails began to angle downward they would scrape the millhouse, crosstrees, or the ground. Many early post mills were damaged in this manner. It was discovered that angling the sail end of the windshaft up from horizontal shifted the thrust forces to the rear bearing, which could be made larger than the front bearing, and tilted the sails farther from the millhouse. From then on practically all windmills for grinding grain were built with angled windshafts.

The next major development in windmills was the tower mill, with a fixed body or tower and a rotatable cap that contained the windshaft. The Dutch introduced the tower mill in the early 1400's. The tower mill represents a significant step forward in the evolution of windmills. It provided a striking contrast to the medieval-appearing post mill, and its use of a fixed tower and a rotatable top cap remains unchanged today in many modern windmill designs. Unlike the post mill, the tower mill could be built to practically any height and could contain as many floors as desired for machinery and storage. In addition, the tower mill did not require an outside ladder which had to be lifted from the ground with the tailpole of the post mill to move the sails into the wind. Finally, the tower mill created the necessity for an invention that would save the miller the work of manually rotating the cap on the curbing of the tower to turn the sails into the wind. This invention, the *fantail,* is a small multi-vane wind wheel set on the back of the cap with its axis at a 90° angle to the windshaft. When the main sails are set directly into the wind, the fantail is shielded from the wind and does not turn. As the wind shifts direction the fantail is powered, and through a set of gears with a ratio of several thousand to one, drives the cap around until the main sails once again face the wind. At this point no wind strikes the fantail and it stops turning until the wind direction changes again. The fantail is a logical and smooth-working appendage to the tower mill. However, fantails mounted to the wheeled tailpole of post mills appear awkward and must be very large because they are set low to the ground where the wind speed is the lowest.

Fig. 1–3: An original sail mill, North Falmouth, MA. (Society for the Preservation of New England Antiquities)

WINDMILL SAILS

Both post and tower mills provided the proving ground for the development of improved windmill sails—the rotating blades that extract energy from the wind. The earliest sails looked like a large paddle (Fig. 1–4A). Each arm, or stock, was set in an opening in the poll end of the windshaft with wedges, and wooden strips called bars were mortised in the stocks, with about equal lengths on each side. Lengthwise wood strips, called *hemlaths,* braced the bars and the sail cloth that was laced in and out of the bars on each side of the stock. The sail cloth was tied to the inner and outer bar of each stock to hold it in place when the blades revolved in the wind. To adjust the speed of the windmill to the prevailing wind speed the amount of sail cloth on each side of the stock was reduced, or reefed, by furling the sail cloth inward toward the stock.

The next major development in windmill sails was truly a giant step by people having little or no understanding of modern aerodynamics. With wisdom that predated science, a windmill sail was developed that obeyed many of the rules of correct airfoil design. Airfoils were unknown at the time, and their real value was not verified until modern wind tunnels became available. In this sail design, which is still in use today, the bars protrude far less on the leading edge and are covered with thin wood strips, called *leading boards* (Fig. 1–4B). A rounded strip forms the leading edge. Sail cloth covered only the trailing bars, uplongs, and the hemlath.

As windmills grew in size and stood taller, rigging cloth sails became an increasingly difficult, time-consuming, and dangerous job. Inventors in the 1700's produced a number of alternatives to common cloth sails, some readily forgettable, while others are still encountered on surviving windmills. Apparently these sails gave good service. These inventions are broadly classified as "patent sails" because their creators scrambled to the patent office to protect and license each device, whether or not it was a practical success.

The most common patent sails were spring-loaded shutters. The spaces between the trailing bars were filled with hinged shutters of wood, or cloth stretched on a wooden frame, each about one foot wide. With the shutters closed the blade is a continuous flat surface to the wind. With the shutters open, like a venetian blind, most of the wind is spilled through the blade. The shutters on each sail are controlled by an adjustable leaf spring and rod mechanism that balances the wind pressure against the closed shutters to the tension set on the spring. When the wind pressure on the closed shutters overcomes the spring tension, the shutters open, spilling the wind. Although this invention was a boon to millers, it still required stopping the mill to adjust the spring tension. Other mechanisms were developed to control the shutters from inside the cap of tower mills while the blades were turning by using mechanical linkages and a hollow windshaft. But by this time the steam engine had been invented, the industrial revolution had started, and interest in working windmills would soon come to a halt.

Other types of shutters include *air brakes* (Fig. 1–4C) and the *sky scraper* (Fig. 1–4D), both of which open into the airstream cut by the blades to slow down the mill. Patent sails never completely replaced the common cloth sails, as they yielded less power from the wind, and the complex mechanisms were trouble prone, especially in wet weather. The Dutch, in particular, disdained the use of patent sails for most of their windmills. Windmills with four blades sometimes used two patent sails and two common cloth sails as a compromise between maximum power and speed regulation.

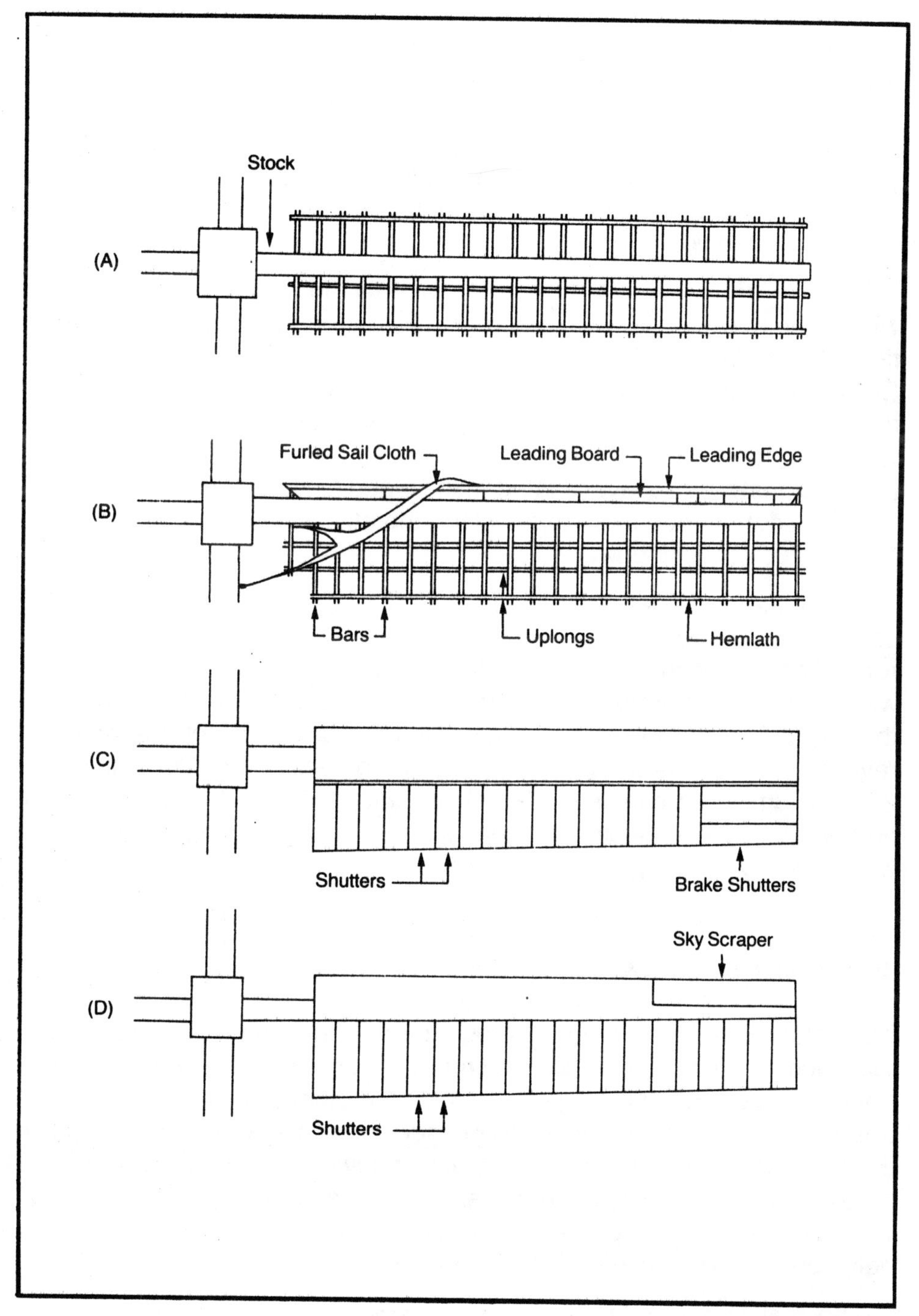

Fig. 1–4: Windmill sails

Most post and tower mills used four blades, although some mills were built with five, six, or eight blades. More than four blades increased the power available to the grindstones, but increased the weight and stress placed on the windshaft. Windmills with an odd number of blades, such as five, became badly unbalanced if one blade broke and could not be run until repairs were made. With an equal number of blades, the blade opposite a broken sail could be removed to balance the mill, allowing the mill to run (with reduced power) until repairs could be made. In the waning days of windmills in England many old mills were running with only two sails—the others being broken at some time and not repaired.

Windmill blades rotate clockwise or counter-clockwise depending on which way the leading edge of each sail faces (Fig. 1–5). Blades were angled, or pitched, into the wind about 15° to 20°. Later the blades were given a twist so the pitch would vary over the length of the blade, being the greatest inboard near the windshaft and least at the tip. These were called *weathered sails,* and angled from about 20° at the inboard edge to about 5° or less at the tip.

Weathering of windmill sails was urged by John Smeaton of England, who in the mid-1700's was among the first to conduct scientific investigations of windmills. His paper on windmill sails was read to the Royal Society in 1759. Smeaton told the gathering that the power available from the wind is proportional to the cube of the wind speed. That is, when the wind speed doubles, the power available in the wind increases eight times. Smeaton also investigated many types of Dutch and English windmills and determined their horsepower and efficiency. His work remains a classic study of wind power.

But neither Smeaton's work nor the work of others to improve windmills could forestall the downfall of the working windmill in England and Europe. For nearly three centuries the Dutch used approximately 10,000 windmills to grind grain, pump water, make paper, and saw lumber. With the introduction of the steam engine and the beginning of the industrial revolution the use of windmills began a sharp decline. By 1900 only about 2500 windmills were in use by the Dutch, with a similar decline in windmill numbers in other industrialized nations. (The Dutch now have only 900 windmills of various types in operation.)

THE AMERICAN EXPERIENCE

The colonists of North America brought with them the knowledge of windmills and lost little time in harnessing the energy from the wind. The first post mill was erected in Virginia in 1620, and others followed along Chesapeake Bay by 1635. Within a few decades windmills spread up and down the Atlantic Coast, but as inland travel and development expanded, watermills along the numerous streams and rivers began to replace the coastal windmills for grinding grain. By the late 1700's few new windmills were being built, and those left standing soon fell into disrepair.

The real boom in American windmills would not come until the late

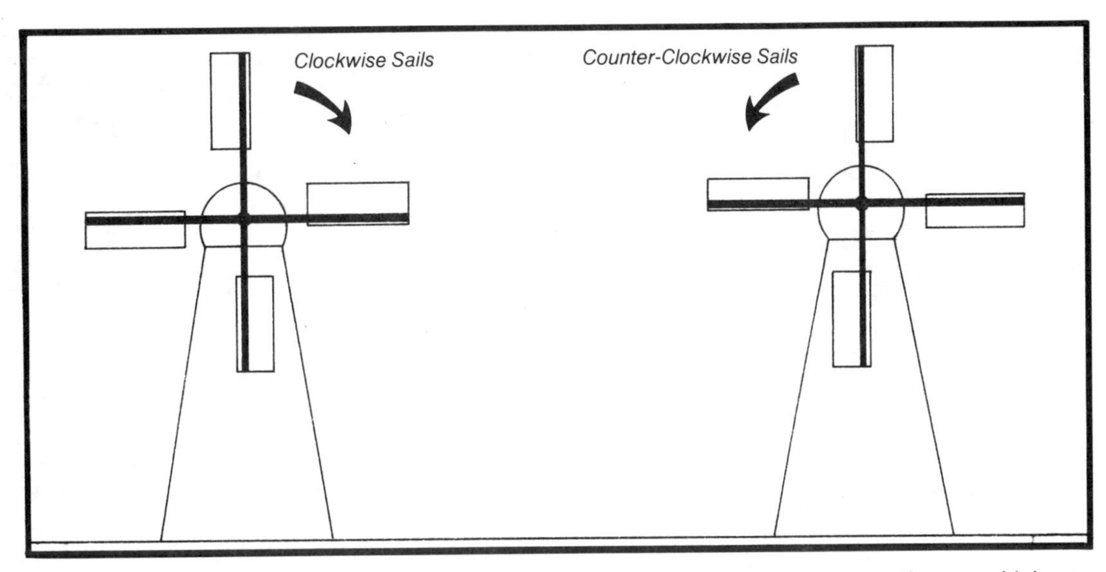

Fig. 1–5: Windmill blades rotate in a direction dependent on which way the leading edge of each sail faces.

The oldest windmill on Cape Cod. (Society for the Preservation of New England Antiquities)

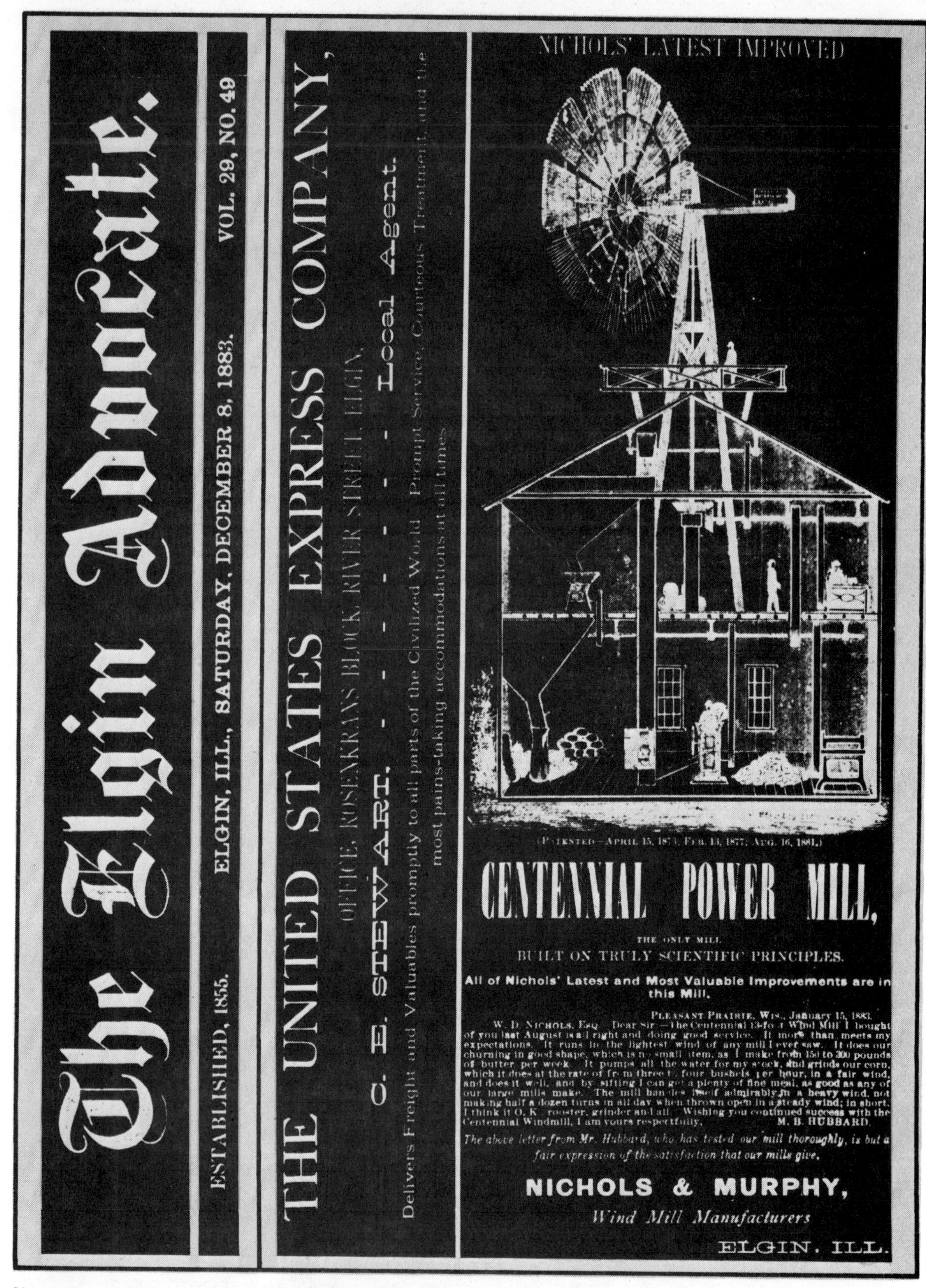

The Elgin Advocate.

ESTABLISHED, 1855. ELGIN, ILL., SATURDAY, DECEMBER 8, 1883. VOL. 29, NO. 49

THE UNITED STATES EXPRESS COMPANY,

OFFICE, ROSENKRANS BLOCK, RIVER STREET, ELGIN.

C. E. STEWART, - - - - - - Local Agent.

Delivers Freight and Valuables promptly to all parts of the Civilized World. Prompt Service, Courteous Treatment, and the most pains-taking accommodations at all times.

NICHOLS' LATEST IMPROVED

(PATENTED—APRIL 15, 187[illegible]; FEB. 13, 1877; AUG. 16, 1881.)

CENTENNIAL POWER MILL,

THE ONLY MILL

BUILT ON TRULY SCIENTIFIC PRINCIPLES.

All of Nichols' Latest and Most Valuable Improvements are in this Mill.

PLEASANT PRAIRIE, WIS., January 15, 1883.

W. D. NICHOLS, ESQ.—Dear Sir:—The Centennial 13-foot Wind Mill I bought of you last August is all right and doing good service. It more than meets my expectations. It runs in the lightest wind of any mill I ever saw. It does our churning in good shape, which is no small item, as I make from 150 to 200 pounds of butter per week. It pumps all the water for my stock, and grinds our corn, which it does at the rate of from three to four bushels per hour, in a fair wind, and does it well, and by sifting I can get a plenty of fine meal, as good as any of our large mills make. The mill handles itself admirably in a heavy wind, not making half a dozen turns in all day when thrown open in a steady wind; in short, I think it O. K., rooster, grinder and all. Wishing you continued success with the Centennial Windmill, I am yours respectfully, M. B. HUBBARD.

The above letter from Mr. Hubbard, who has tested our mill thoroughly, is but a fair expression of the satisfaction that our mills give.

NICHOLS & MURPHY,

Wind Mill Manufacturers

ELGIN. ILL.

Newspaper ads like this appeared regularly in the 1800s.

Headquarters of the Aermotor Company, Chicago, Illinois, as it appeared in the 1800s.

THE BARGAIN COUNTER.

Under this heading, in each Aermotor Bulletin, we shall offer some useful article to the hardware, implement and windmill trade at a price so low as to be practically regardless of cost. It is done with the avowed purpose of making dealers look for the Bulletin with "an eye to business."

With this issue we offer the Aermotor Crab at $5.00. The first one that the Aermotor Company had they bought second-hand at $20.00, and have since listed and sold them at $20.00. We now offer this Crab made in improved form at $5.00, simply to show what we can do.

PRICE $20.00

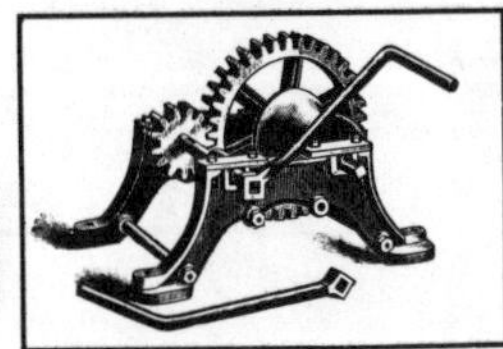

REDUCED TO $5.00

THIS PRICE DOES NOT HOLD GOOD AFTER NOVEMBER 1st, 1893.

For quickly, safely and easily raising windmill towers it has no equal. It does away with the uncertainties of raising windmills with a team. With the Aermotor Crab you do not have to call in all the neighbors to raise a windmill. For an ordinary Pumping Outfit, if suitable tackle blocks are used, two men, one to turn the crank and another to hold the rope, are all that are necessary.

Orders received during October will be filled at this price, and filled in the order in which they are received, cash to accompany the order. Orders not accompanied by the cash will not be acknowledged, and will have no attention paid to them. Communications received after November 1st, in regard to this special offer will not be replied to. Any one wishing to take advantage of this offer should DO IT NOW, while it is good.

LOOK FOR OUR NEXT BARGAIN.

BARGAIN COUNTER.

We now place on the bargain counter, a Lift Pump Standard, with the latest improvements, well finished, tapped for two-inch pipe, bushed for smaller pipe, if desired, with a ten-inch stroke and most desirable from every point of view, at about two cents a pound, or much less than one-half of a rock-bottom price for such a Pump Standard.

Whether or not this is a bargain, and whether or not we can do it at a profit, we will leave to those who are well informed. We wish to make only these stipulations. That you order not to exceed one dozen, and that the order shall get here not later than December 1st; and that if this should chance to fall into the hands of any who are not our regular agents, who have not a credit established with us, and who are not rated in the commercial agencies, so as to warrant us in shipping to them without further inquiry, that they enclose cash with order.

These Pump Standards are suitable for windmill or hand work. You will understand that this does not include cylinder or pipe, but only what is seen in the cut.

Cash, with order, $1.55 each. Orders not accompanied by cash, $1.75. For orders received after December 1st, the price will invariably be $4.58.

CUT FROM $4.58 DOWN TO $1.55.

This advertisement appeared in an 1893 edition of the Aermotor Co. house organ.

Fig. 1–6: An early Aermotor multi-vaned windmill. (Photo by Tom Kovarik)

1860's, when the distinctive American water-pumping windmills were developed and manufactured in the tens-of-thousands. The greatest impact on the history of windmills that these water-pumpers had was to usher in the era of the factory-built wind machine that could be erected by practically anyone. Factory production would mean low cost, standardized parts, and widespread distribution by wandering windmill salesmen and orders through mail-order catalogs.

The market was huge. Railroads cutting across the continent needed countless windmills to fill the water tanks of steam locomotives, and sodbusters moving westward along the rail lines needed windmills for household water and for livestock. The American water-pumping windmill helped bring running water and flush toilets into the average home in rural areas and small towns without central water systems. The wonderful windmill could pump water to the storage tanks on the upper floors of the tall Victorian houses of this period.

By 1890 the nearly four-score windmill factories in the United States began to look to foreign markets, and American water-pumpers sprouted practically everywhere from Australia to Yucatan. By the turn of the century most of the windmills were of all-metal construction, with multi-vaned wind wheels, or fan-blades, of 12 to 16 feet in diameter (Fig. 1–6). Most used a tail-vane to keep the blades facing into the wind. With water-pumpers, the windshaft is connected to a set of gears and a cam that moves a connecting rod up and down. This, in turn, operates a pump at the base of the tower. These are slow-speed machines, but in a 15 mph (24 kmph) wind a 12-foot (4-meter) wind wheel could pump 35 gallons (133 liters) of water a minute to a height of 25 feet (8 meters).

The golden era of the American water-pumpers ended in the 1930's, when over 6,000,000 were in operation. The machines were so durable that once a farmer bought one he would never have to replace it. When demand slackened, many manufacturers passed into oblivion. The final deathblow was brought by rural electrification, when electric motors replaced wind power. Today, there are very few factories in the United States producing water-pumpers, and most are exported.

THE FIRST FLICKER OF LIGHT FROM THE WIND

Just as the American water-pumpers improved life by bringing running water into countless homes, the first small wind generators for producing electricity helped relieve the boredom that isolation and darkness brought to America's rural areas. When the boys came home after the World War I they knew the cheerfulness a few electric lights could bring. As the first crystal sets and tube radios appeared, the rush was on to build small wind generators that could charge a couple of batteries. Generators were removed from old cars and connected to the multi-vaned wind wheels from water-pumpers, usually with disappointing results. Before long backyard tinkerers realized that the slow speed wind wheel of water-pumpers didn't turn fast enough to produce electricity from a generator.

What was needed was a fast-turning propeller like those attached to the engines of the cloth and wood biplanes that flew above the trenches during the war. Only a few lucky fellows could find old airplane propellers, but farm boys throughout the Midwest and the Great Plains carved imitations from sawmill slabs. So many people wanted a small wind generator to light a few small lamps and operate a radio that the experimental stations of practically every land-grant state college published a booklet or a circular on how to construct a small home-built wind-electric plant. These plants were as simple as a hand-carved propeller attached to a Model T generator on one end of a wooden board, with a tail-vane attached to the other end. The board was pivoted on an iron pipe and flange, which slipped into another pipe fastened to the top of an old water-pumper windmill tower. No one knows how many of these plants worked, but those that did increased their amateur builders' desire for even more power. Soon some of the water-pumper windmill factories and newly formed companies began to market factory-built wind-generator plants. Many were so bad that they were forgotten as soon as the companies went out of business. A few wind generators, such as the Jacobs, were so good that their memory lives on among windmill cultists.

Marcellus Jacobs began his experiments with wind generators in the early 1920's when a small wind generator at his father's ranch in Montana did not deliver enough power to be useful. He built his first wind generator from a water-pumper wind wheel but soon realized that this slow-speed windmill fan produced very little power. In 1927 Jacobs finished the development work that led to an efficient three-blade propeller and a smooth-working flyball governor. He combined a 15-foot (4.5-meter) propeller and his patented governor with a massive 440-pound (199.5-kilogram) battery-charging generator of his own design. His 32-volt generator was rated at 2500 watts, and the 110-volt generator could produce 3000 watts. Jacobs sold the 32-volt generator for $490. He also sold a 50-foot (15-meter) steel tower for $175, and a 21,000 watt-hour glass cell lead-acid storage battery set for $365. The tower could be built and the generator installed by two men in two days.

Between 1931 and 1957 thousands of Jacobs plants were sold and installed in all parts of the world, including weather stations within the Arctic Circle and at Little America in Antarctica. In 1936 Jacobs designed a special wind generator for the cathodic protection of underground steel pipelines. Hundreds of these plants were used to protect pipelines in North and South America and in Arabia.

Jacobs stopped building new generators in the 1950's when the business was no longer profitable. By this time, all but one of his competitors had stopped making wind generators. This company, the Winco Division of Dyna Technology, Inc., of Sioux City, Iowa, still manufactures the 12-volt Wincharger wind electric battery charging generator (Fig. 1–7). The Wincharger is a small plant, with a 6-foot (2-meter) two-bladed wood propeller, and is sold as a complete unit with a patented air-brake, a wired instrument panel, and a 10-foot (3-meter) steel tower. With a maximum output of only 14 amperes in a 23 mph (37 kmph) wind, the Wincharger is designed primarily for emergency or standby lighting and communications equipment and is not really suited for supplying power to the average home.

Fig. 1–7: The 12-volt Wincharger generator. (Winco Division of Dyna Technology)

AMERICA'S LARGEST WIND GENERATOR

In 1934 Palmer Cosslett Putnam, a young engineer, aviator, and yachtsman from an influential East Coast family, built a home on Cape Cod and discovered that both the wind speed and the electric power rates were surprisingly high. Putnam reasoned that a windmill to generate electricity could take advantage of the strong winds and reduce his power costs. But the small battery-charging wind generators that produced direct current were not what Putnam had in mind. He wanted a windmill that would generate alternating current, the same type of electricity that the power company supplied. His bold plan included having the power company maintain stand-by service should the wind fall to low speeds. When the wind was especially strong, his windmill would feed back the excess alternating current into the lines of the power company.

Putnam's plan to combine his windmill with the power company lines is widely endorsed today, but when he calculated the size of a windmill that could supply the peak load of his all-electric home he discovered that the wind generator would have to be much larger than the small battery-chargers then on the market. Putnam studied different types and sizes of wind generators and concluded that to be economically sound his alternating-current wind generator must be made so large that it should be owned and operated by the power company and connected to the company-owned grid, or to a distribution network of power lines. Only as part of an existing power system with steam or hydroelectric generating stations could the alternating-current wind generator be a success. When the wind was strong, a bank of wind generators could supply most of the power to the grid; when the wind was weak, the hydroelectric or steam generators would take up more of the load.

Putnam organized a group of scientists and engineers to assist him in the design and placement of a large wind generator. After much preliminary design work he called upon the captains of industry to sponsor his project. Thomas S. Knight, a fellow yachtsman and vice president of General Electric, became involved and introduced Putnam to the S. Morgan Smith Company, who agreed to develop and test Putnam's ideas. In 1939 Putnam's team, who were by then a veritable *Who's Who* of American science, engineering, and industry, signed on the Central Vermont Public Service Company to be the first electric utility in the United States with an alternating-current wind generator connected to its power grid.

While site selection and engineering tests were in progress early in 1940, it became clear that the United States would soon become involved in the war that already raged in Europe. When this happened, materials and manufacturing companies would no longer be available to such a fanciful project, so the team decided, early in the spring of 1940, to freeze the present test design, order parts and forgings, and choose the site for the massive wind generator. The site chosen was Grandpa's Knob, a 2000-foot peak in the Green Mountains of central Vermont, 12 miles from the town of Rutland.

The Smith-Putnam wind turbine, as it became known, was a real giant. Its two glistening stainless steel blades spanned 175 feet (53 meters), perched on a 110-foot (33.5-meter) tower. The unit weighed 250 tons (227 tonnes) and could generate 1250 kilowatts, enough for a small town. A road to the site was built during the summer and fall, with tower construction continuing into the frigid winter. Assembly of the wind generator started in late winter and was completed in August, 1941. The blades were rotated by the wind for the first time on August 29, 1941. At 6:56 PM on October 19, 1941 the generator was engaged, and electricity was delivered to the power company grid.

A test program was conducted to March 3, 1945, when the wind generator was turned over to the power company for routine operation as a generating station. During the test program the wind generator was operated for approximately 1100 hours. There were many interruptions caused by mechanical failures and for modifications and refinements to achieve smooth running. The longest downtime, lasting 25 months, began in February, 1943, when a main bearing failed. Wartime priorities and material shortages greatly lengthened these interruptions, and most of the original design and engineering team left for more important war work. At 3:10 AM on March 26, 1945, during routine operation by the Central Vermont Public Service Company, one of the blades broke loose and fell to the ground, ending America's first experience with a large wind generator.

The Smith-Putnam wind turbine was disappointing to its builders only because it was rushed into completion when the storm clouds of war threatened. Due to insufficient wind surveys the wind at Grandpa's Knob was only about a third of what was estimated. And there was not sufficient time to conduct stress tests of model wind turbines, which would have revealed the weak points at the roots of the windmill blades. But the project stands alone as a great effort by some of America's greatest scientists and engineers of the 1940's to harness the wind.

WIND POWER DEVELOPMENTS IN OTHER COUNTRIES

In the 1930's other countries were also working to exploit wind energy. The Russians built a 100-foot (30.5-meter) diameter wind generator in 1931 near Yalta, overlooking the Black Sea. This machine generated alternating current and was connected to a power grid also supplied by a steam generating plant. The Russian design was bold yet simple, using handmade wood gears and blades covered with roofing metal. Putnam was impressed by the sturdiness of the Russian design but thought that the United States could do much better. In 1933 a German, Professor Hermann Honnef, suggested a design utilizing five 250-foot (76-meter) diameter rotors atop a 1000-foot (305-meter) tower. While this design has been attributed to Nazi propaganda, with construction never seriously considered, smaller but more practical wind generators were at work in other countries.

Denmark

During the last decades of the nineteenth century agriculture was rapidly mechanized, and practically every Danish farm had a small windmill atop the barn roof that powered threshing or milling machinery through a mechanical drive. These windmills, which numbered about 30,000 were of primitive construction, made of wood and fitted with cloth sails. In 1891 Professor Poul la Cour, a teacher of physics and chemistry, became interested in windmills that could generate electricity. The Danish Government sponsored his work, and he was able to construct a scientific laboratory and wind tunnel for studying model windmills. La Cour built a number of wind generators to supply direct current to his village and school. By 1910 several hundred systems designed by Professor la Cour were built and in operation. These consisted of 80-foot (24-meter) towers that supported a 75-foot (23.7-meter) diameter four-bladed wind rotor that operated a gear and shaft connected to a generator on the ground. These generators were rated at from 5 to 25 kilowatts.

During World Wars I and II improved wind generators of this type supplied much of the Danish power needs, as the supply of oil into the country was blocked. After World War II the Danes developed and operated three experimental wind generators, rated at 12, 45, and 200 kilowatts. These were alternating current wind generators and were connected to electric power grids.

In 1967 the 200 kilowatt Gedser wind generator was shut down because operating and maintenance costs were too high in comparison to costs at steam generating plants. In 1974 the Danes restudied wind generator costs and concluded that as energy costs continue to rise wind generators would soon become economically feasible once again.

England

During the late 1940's and into the next decade considerable work was done on wind generators in the British Isles. Wind measurements made at about 100 sites during this period are among the most thorough on wind characteristics for a particular geographical area. In 1950 the North Scotland Hydroelectric Board commissioned the John Brown Company to construct a large wind generator on Cape Costa in the Orkney Islands. This unit was designed to generate 100 kilowatts of alternating current in a 35 mph (56 kmph) wind. A second 100 kilowatt machine was scheduled for erection in North Wales, but, interestingly, the local population was strongly against a wind generator installation. A public inquiry ruled in favor of the local inhabitants, and after preliminary tests at a site about 25 miles (40 kilometers) from London this wind generator was dismantled, shipped to Algeria, and erected on a hill-top site.

In 1953 a windmill research station was organized at Cranfield to conduct wind generator experiments and to develop instrumentation and testing methods for wind generators. In 1958 a third 100 kilowatt wind generator was constructed on the Isle of Man, and a small 8 kilowatt wind generator was installed as a research project to supply power to an isolated homestead. In 1962 nearly all government sponsored wind research projects came to an end. Today, small scale wind generator development and testing is in progress at many universities and technical colleges.

France

At the end of World War II work began in France to develop large wind generators that could deliver alternating current to electric utility distribution grids. After many years of development and testing models in wind tunnels the first large machine, a three-bladed wind generator, was erected near Paris in 1957. This experimental machine weighed 160 tons (145 tonnes), and its 100-foot (30.5-meter) rotor was connected to a generator that could deliver 800 kilowatts to the power grid when the wind speed was 36 mph (58 kmph) or higher. Between 1958 and 1962 this wind generator operated for more than 5000 hours and was connected to the the power grid for about 600 hours. During the test program a set of newly designed flexible blades was installed, but these proved unsound, and one blade broke, destroying the hub of the machine. This wind generator was dismantled in 1965.

Two other wind generators were constructed in Southern France in 1963. The smaller unit had a 70-foot (21-meter) rotor and was rated at 132 kilowatts in a wind of 28 mph (45 kmph). The larger machine was rated at 1000 kilowatts in a wind of 37 mph (60 kmph) and had a rotor nearly 120 feet (21 meters) in diameter. Early in 1964 a mechanical failure stopped this larger machine, which was dismantled instead of being rebuilt. The Bureau of the French Electricity Authority, the organization conducting wind generator research and development, was dissolved in 1966, and many of the key engineering personnel joined the newly formed Aerowatt Company, a private firm that continues to manufacture small wind generators for the world market.

RECENT WIND POWER DEVELOPMENTS IN THE UNITED STATES

Federally sponsored work to harness wind energy within the United States began in 1973 as part of the nation's solar energy program. The first large wind generator project was assigned to the NASA Lewis Research Center by the National Science Foundation, which managed the wind energy program. This program recommended the design, construction, and testing of a 100-kilowatt, 125-foot (38-meter) diameter wind turbine (Fig. 1–8) as quickly as possible to provide engineering information that could be applied to a large wind energy program. This wind generator was

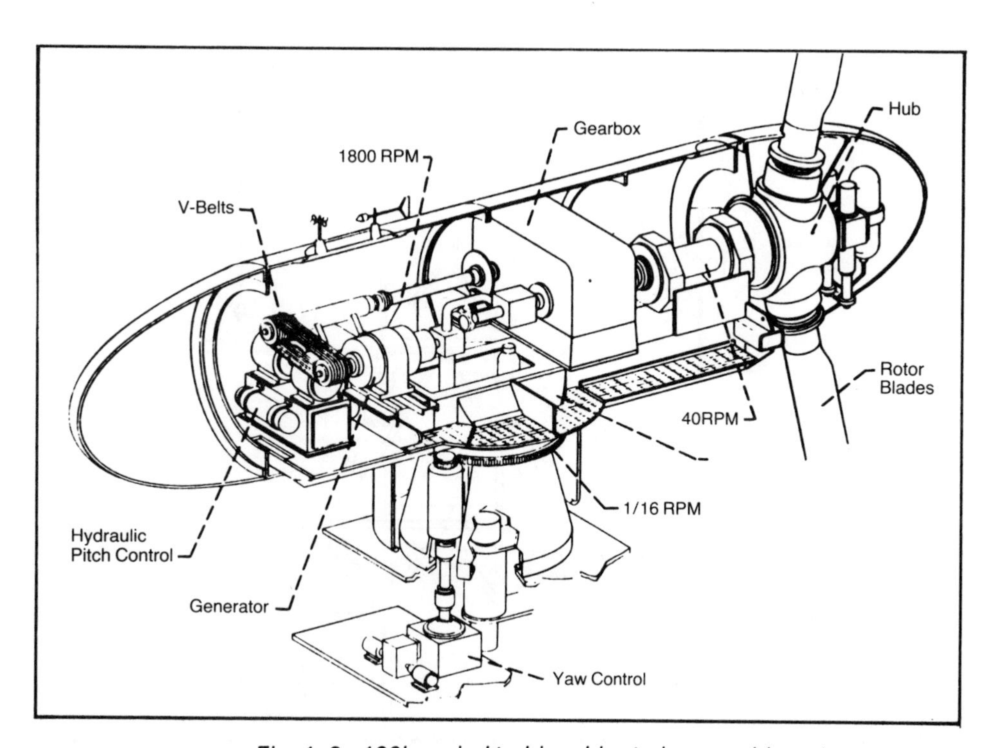

Fig. 1–8: 100kw wind turbine drive train assembly and yaw system.

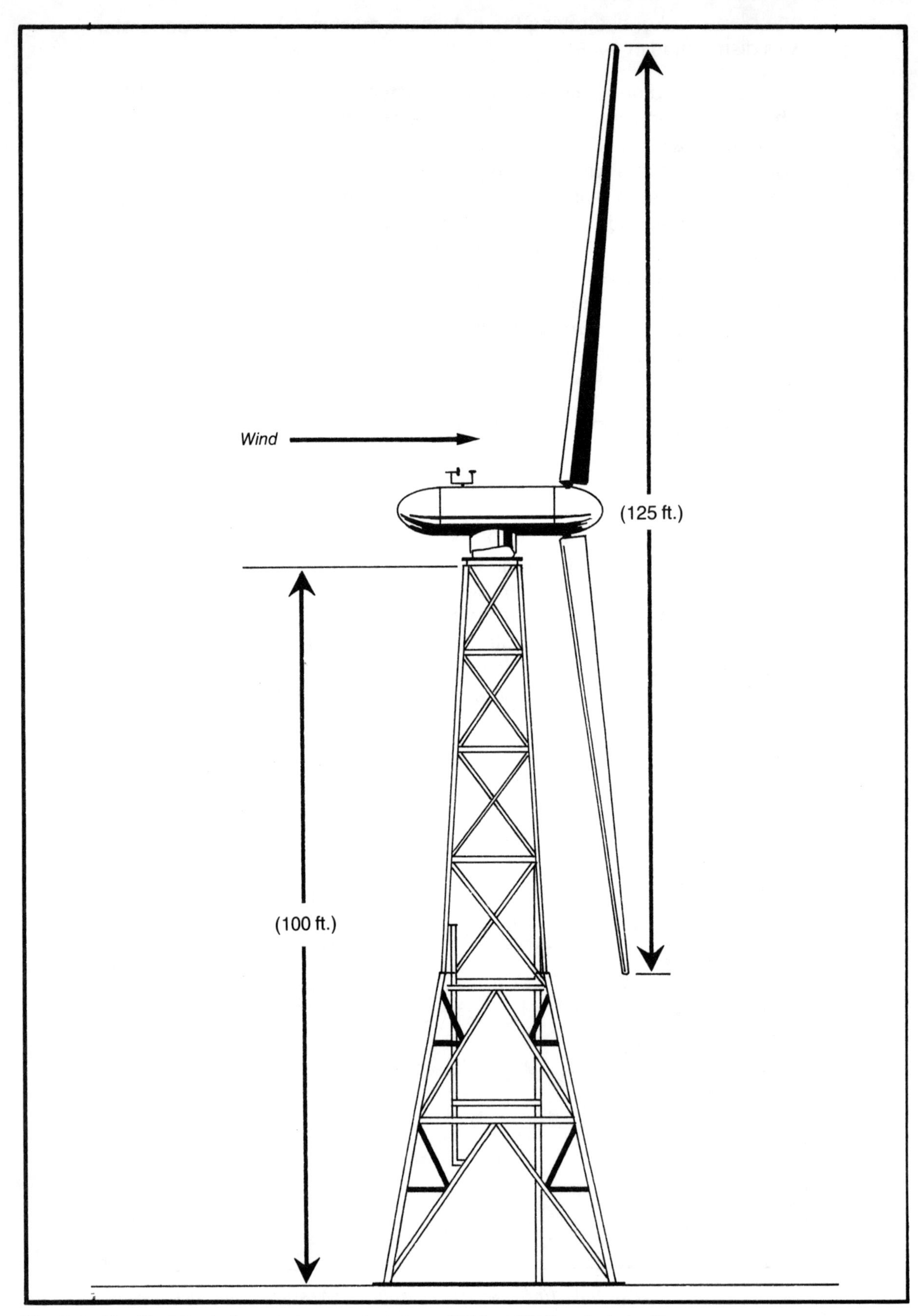

Fig. 1–9: The Mod-O wind generator mounted atop a 100-foot (30.5-meter) steel tower.

designated Mod-O, and would be located at the NASA Plum Brook site near Sandusky, Ohio, where extensive wind records have been compiled for ten years.

In January, 1975, the responsibility for managing the wind energy program was transferred to the newly formed Energy Research and Development Administration (ERDA). This change did not affect Mod-O, which became operational in September, 1975. The wind turbine was designed, built, and erected in 18 months.

This test machine consists of a two-bladed wind rotor, step-up gearbox, generator, connecting shafts, control mechanisms, and the tower. The blades are operated at 40 rpm. In the gearbox this speed is increased to 1800 rpm to drive the generator. The entire assembly (Fig. 1–9) is mounted in a streamlined housing atop a 100-foot (30.5-meter) steel truss tower.

The machine is designed to start generating electricity when the wind speed reaches 8 mph (13 kmph). Its full rated output of 100 kilowatts is produced at a wind speed of 18 mph (29 kmph). Above this speed the blades are gradually feathered to keep the generator speed constant. The blades rotate down-wind of the tower and turn counterclockwise looking upwind.

Description of the Mod-O Wind Turbine

Blades The two blades are of all-metal construction; each blade is 62.5 feet (19 meters) long and weighs approximately 2000 pounds (907 kilograms). The blades were fabricated by the Lockheed Corporation.

Hub The hub connects the blades to the low-speed main shaft, which connects to the stepup gearbox. The hub is fixed rigidly to the main shaft by bolts and allows changes only of the blade pitch by the automatic speed control system.

Generator The generator is an 1800 rpm synchronous alternator supplied by the General Electric Company. It weighs 1425 pounds (646 kilograms) and includes a direct-connected brushless exciter and regulator. This alternator produces 480 volt three-phase current that can be stepped-up or stepped-down to different voltages by using transformers.

Drive Train The low-speed main shaft drives the step-up gear box which is a standard commercial unit rated at 236 horsepower. It weighs 4800 pounds (2177 kilograms). The high speed shaft from the gearbox drives the alternator through a belt system.

Bed Plate and Yaw Control The above units of the wind turbine are supported on a bed plate atop the tower that rests on a circular gear-bearing assembly that turns the entire housing to keep the blades in the windstream.

Blade Pitch Control The pitch, or angle of the blades to the wind, is changed by a hydraulic fluid system. The wind turbine generates approximately 100 kilowatts of electricity at wind speeds of 18 mph (29 kmph) and greater. Between 8 mph (13 kmph) and 18 mph (29 kmph) the amount of electricity generated varies with the wind speed. Above 18 mph (29 kmph) the blade pitch is changed to keep the alternator speed constant.

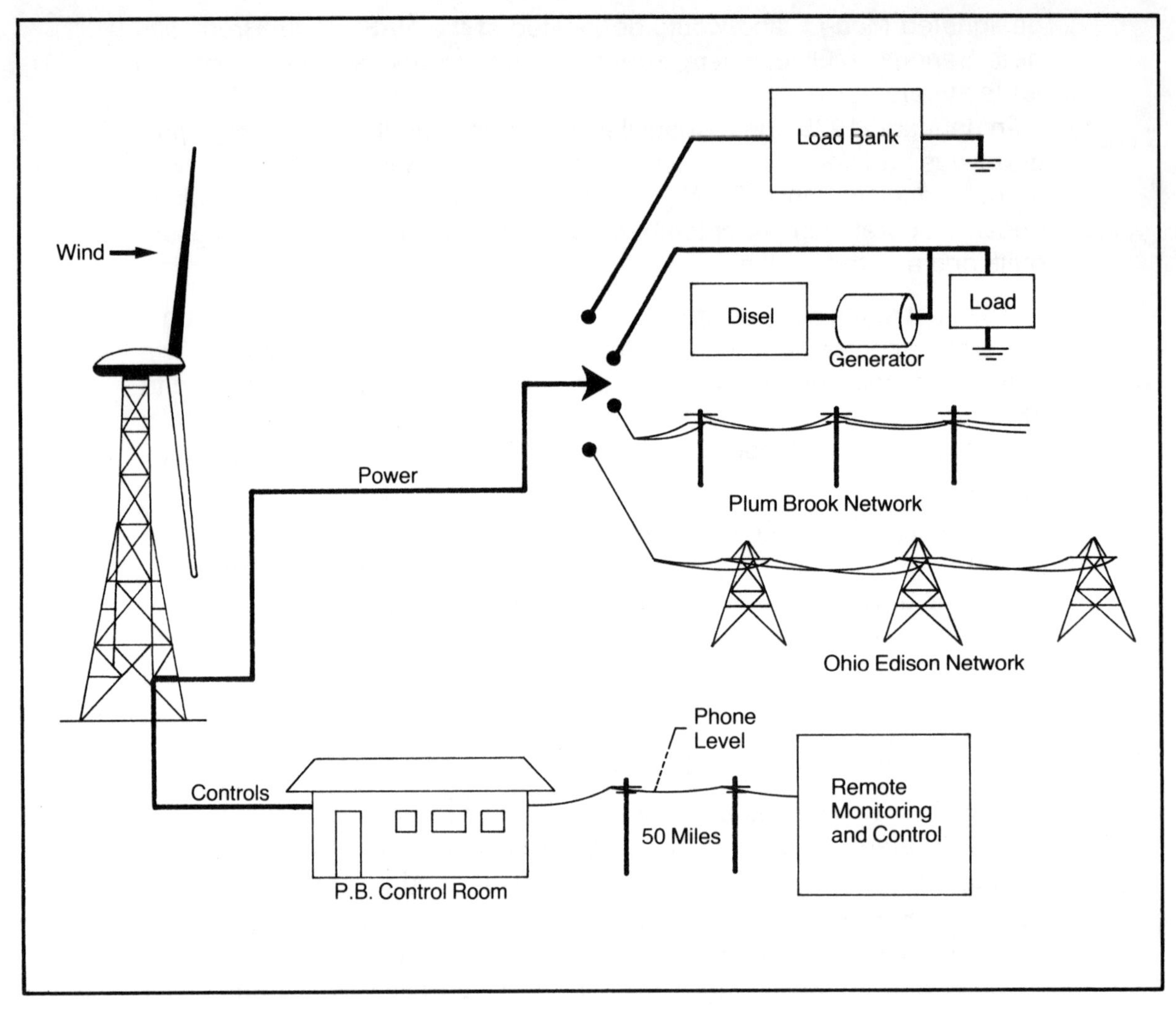

Fig. 1–10: Schematic representation of Mod-O installation at Plumb Brook.

Operation of the Mod-O Wind Turbine

Operation of Mod-O wind turbine consists primarily of startup, normal operations connected to the utility power grid, shutdown, and standby. Fig. 1–10 shows the different modes that Mod-O operates in at Plum Brook. In addition to connection to the Ohio Edison grid, Mod-O can be connected to: 1) a resistive load bank; 2) a diesel generator of approximately 160 kilowatts; and 3) the Plum Brook power network. The Plum Brook network can be disconnected from the Ohio Edison grid to provide a good simulation of a small utility network with several small generators.

The basic wind turbine controls for Mod-O are: 1) the yaw control for aligning the wind turbine with the wind direction; and 2) the blade pitch control used for startup, shutdown, and speed and power control. All normal operating functions are programmed into a computer which provides routine control of the wind turbine. A safety shutdown system is wired into the computer to stop the machine in the event of operating difficulty or a mechanical failure.

Fig. 1–11: Mod-O was first operated at its rated power in 1975.

Performance of the Mod-O Wind Turbine

The 100 kilowatt Mod-O wind turbine was first operated at its rated speed and power in December, 1975 in winds of 25 to 35 mph (40 to 56 kmph) (Fig. 1–11). At this time the machine performed as expected except for some larger than anticipated loads caused by blade bending movements. This bending did not damage the blades, but continuous operation of the test unit with these loads could have resulted in early failure of the blades.

After intensive study it was determined that much of the blade bending movement was caused by the impulse applied to each blade as it passed through the wake of the tower. By performing wind measurements behind the tower and wind tunnel studies, researchers concluded that the tower was blocking the windflow much more than had been expected. To reduce wind blockage caused by the tower the stairways and rails were removed. This modification increased the airflow through the tower by approximately 50 percent. In addition, movements in the yaw system that rotate the wind generator were stiffened, and a fluid coupling was added between the gearbox and the alternator to further smooth power oscillations as the rotating blades passed through the tower wake.

The Mod-O wind turbine has excelled in its role of providing the entire U.S. wind energy program with up-to-date information on the operation

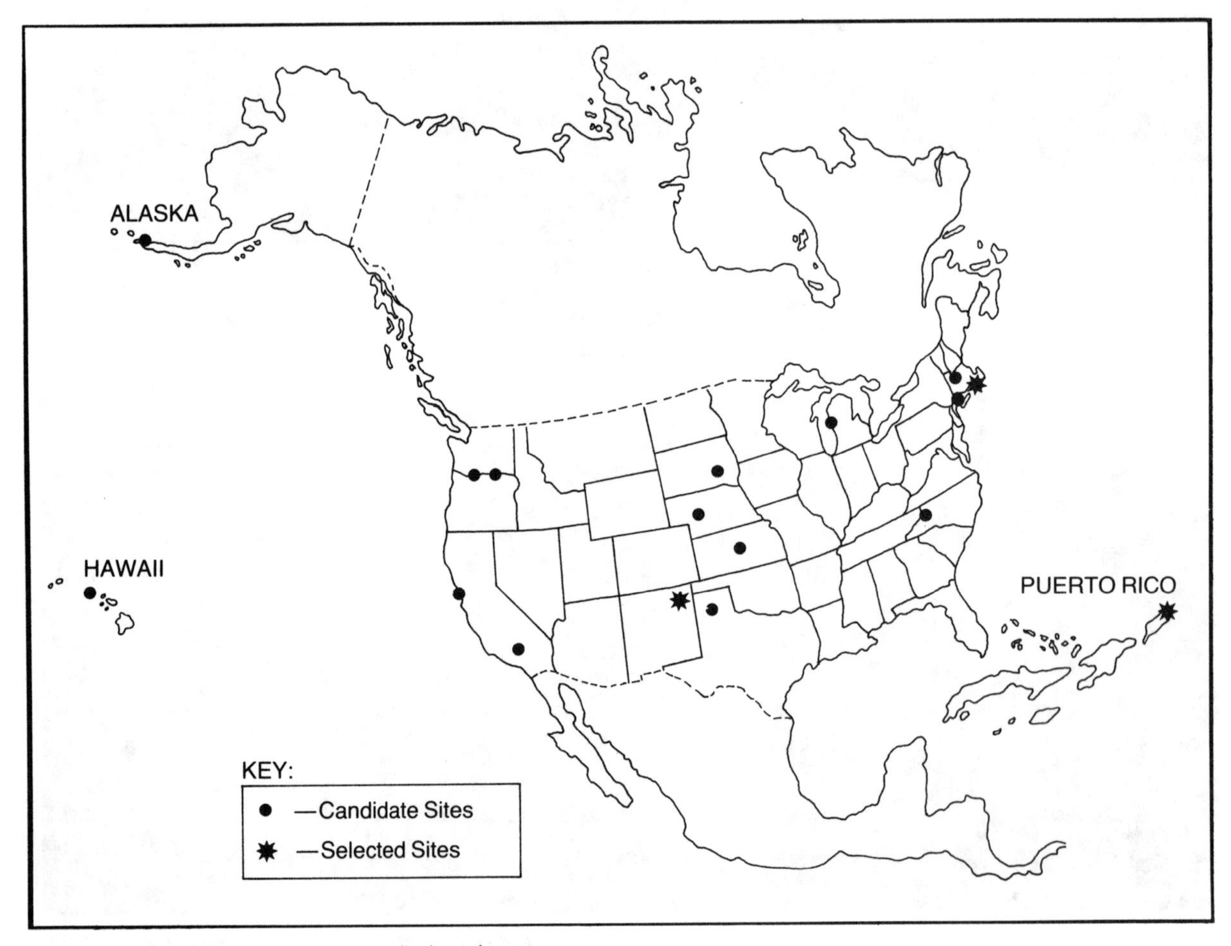

Fig. 1–12: Candidate wind turbine site locations.

and performance of a large, modern wind generator. Mod-O has served as a test bed to develop the latest electronic instrumentation to control wind generators. Operating procedures such as startup, blade pitch changes for the control of speed and power, and synchronization to an electric utility power grid have become routine. The machine has been operated in a fully automatic mode by the use of a remote control and monitor panel located 50 miles (80.5 kilometers) from the wind generator site. This operation simulates how an electric utility dispatcher would control a bank of large wind generators in a future installation. The Mod-O wind generator will also be used to test new ideas in wind generator blade design and fabrication. Many new developments that could reduce the cost of future wind generators will likely be tested first on this versatile machine.

THE MOD-OA WIND TURBINE

In 1975 the Energy Research and Development Administration decided to implement a program of installing large wind turbines at selected sites for the primary purpose of supplying significant amounts of electrical energy into the utility power grid. To accomplish this goal in the shortest length of time, the basic design of the Mod-O wind turbine was utilized but was scaled up to a 200-kilowatt machine. The new wind turbine was designated Mod-OA.

The overall objective of the Mod-OA program was to design, manufacture, and install three 200-kilowatt wind turbines in existing electric utility grid systems. A major part of the program was to collect meterological data at 17 selected sites where the installation of a wind generator would be most beneficial. Three of these sites were selected for Mod-OA wind turbines. The candidate sites and the three sites chosen for wind turbines are shown in Fig. 1–12 and are listed below:

*Block Island, Rhode Island
*Culebra Island, Puerto Rico
Augspurger Mt., Washington
Amarillo, Texas
*Clayton, New Mexico
San Gorgonia Pass, California
Cold Bay, Alaska
Oahu, Hawaii
Ludington, Michigan
Huron, South Dakota
Point Arena, California
Long Island, New York
Boardman, Oregon
Russell, Kansas
Holyoke, Massachusetts
Boone, North Carolina
Kinsley Dam, Nebraska

**Designates sites selected for a Mod-OA wind turbine installation*

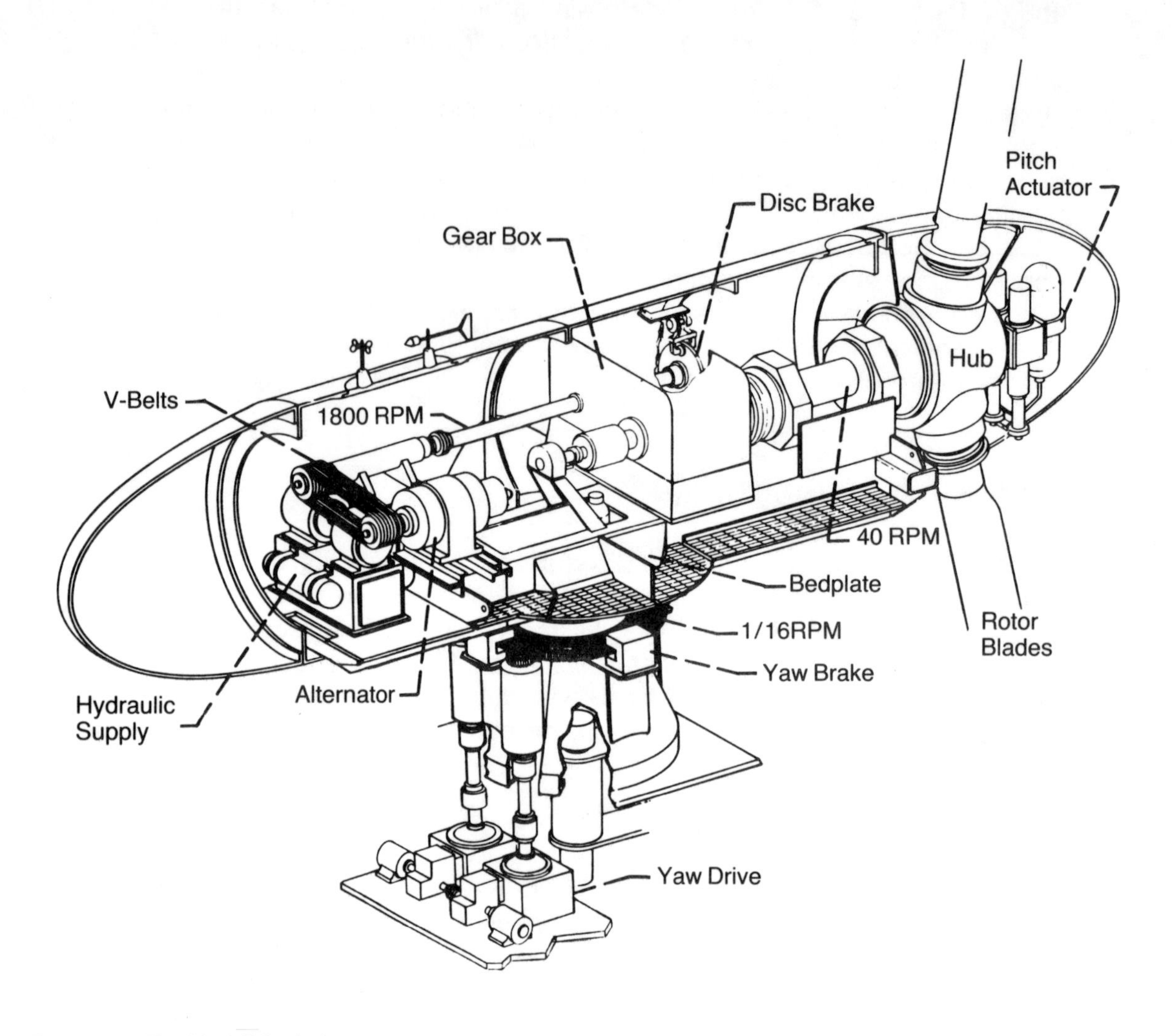

Fig. 1–13: The Mod-OA wind generator.

The 200-kilowatt Mod-OA wind turbine has a two-bladed wind rotor 125 feet (38 meters) in diameter. Mechanically, the Mod-OA machine (Fig. 1–13) is very similar to the Mod-O wind turbine, except for a 200 kilowatt synchronous alternator and longer blades. The rotor is located downwind of the tower and is designed to rotate at a constant 40 rpm when the machine is generating its rated power.

Particular attention was given to interfacing these wind generators with the electric utility power grid. An automatic control system was developed that will start the wind turbine at the correct cut-in wind speed, bring the machine to the design speed, synchronize the alternator with the utility grid, and control the output power level under varying wind conditions. Should the wind speed drop to a low level the wind turbine will be automatically shut down until the wind speed increases to the required start-up speed. The first Mod-OA wind turbine was assembled at Clayton, New Mexico, during November, 1977.

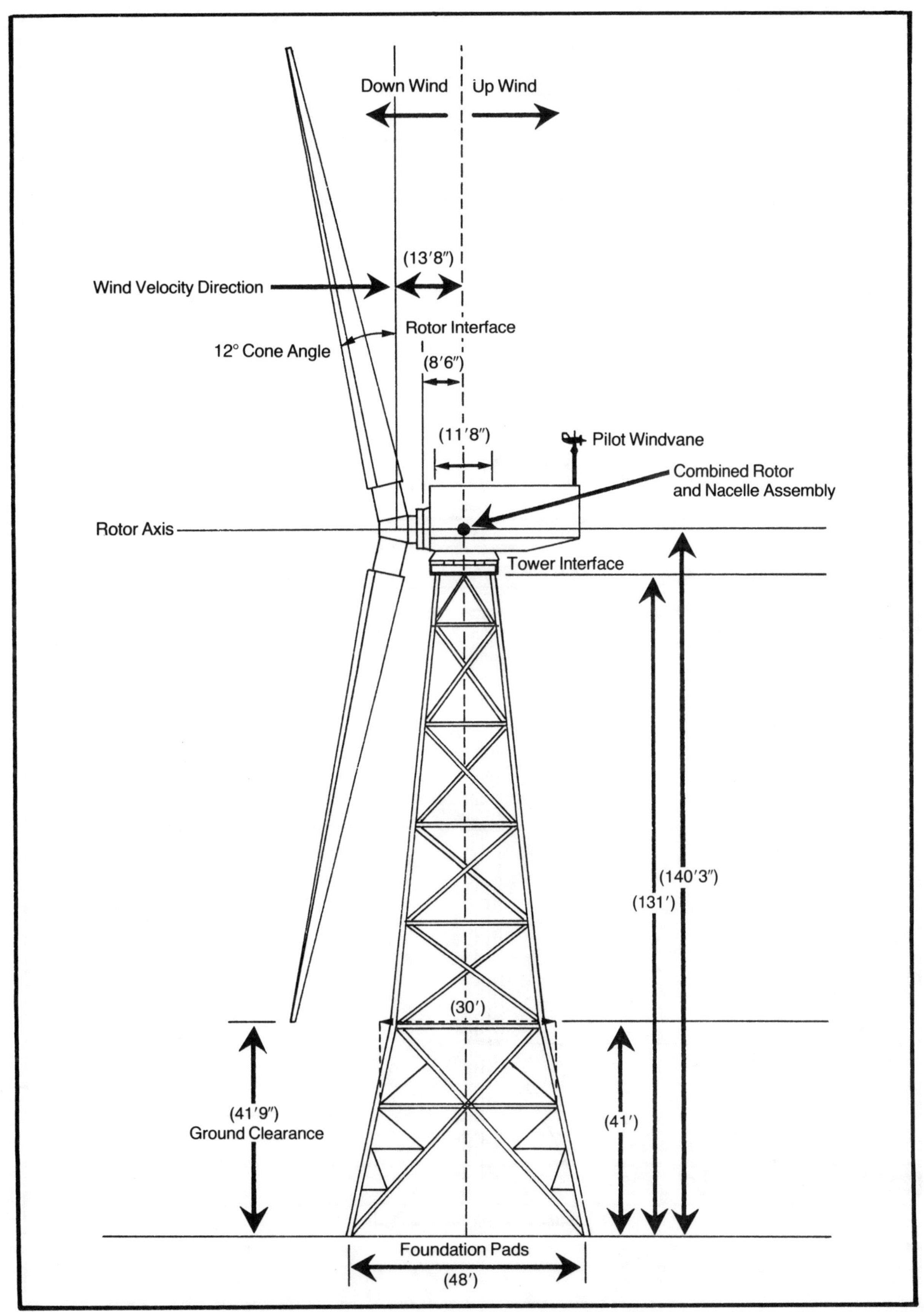

Fig. 1–14: The giant Mod-1 has a rated electrical output of 2000 kilowatts in a 25 mph (40 kmph) wind.

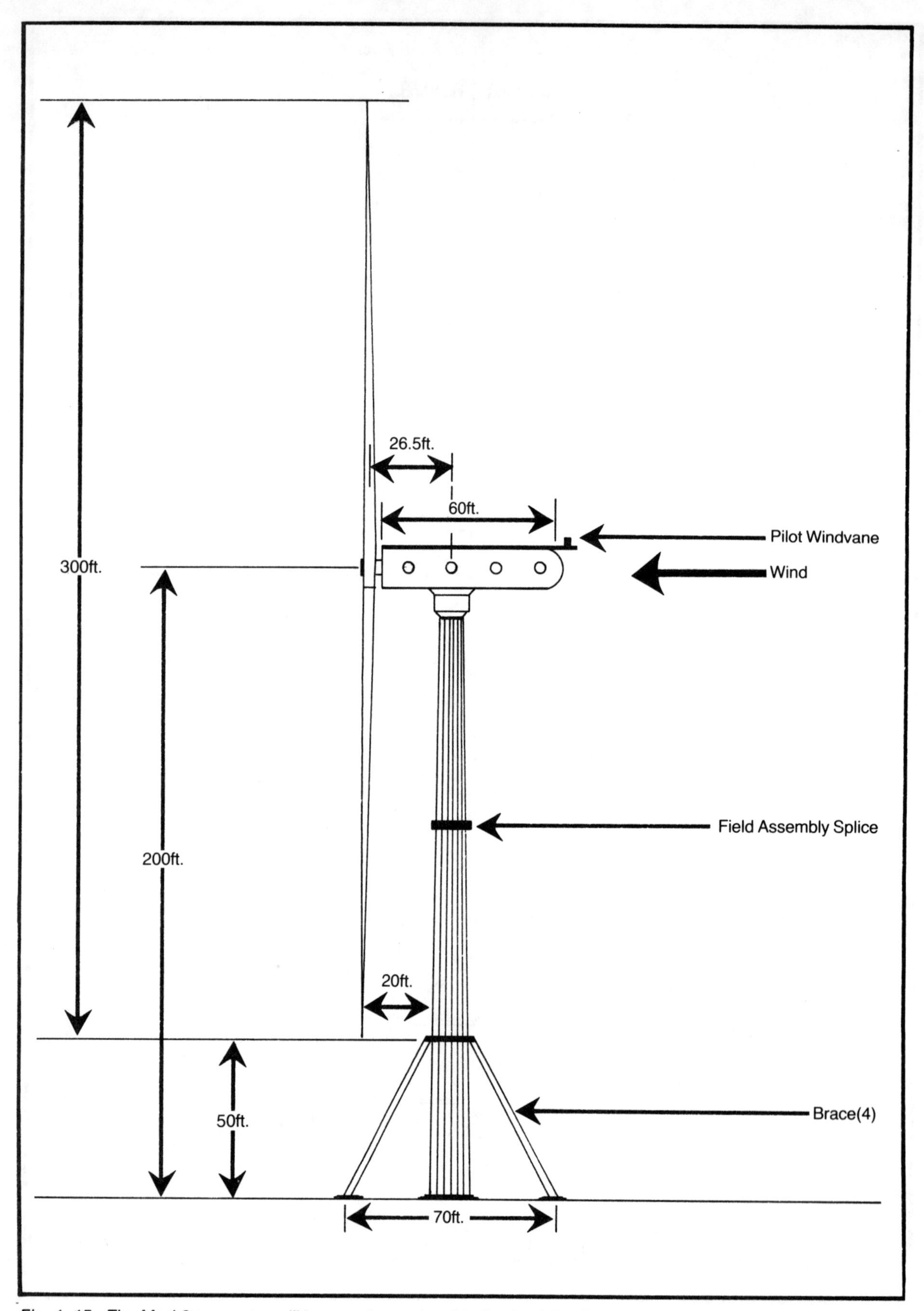

Fig. 1–15: The Mod-2 generator will be even larger than Mod-1, but it is still in the planning and design stages.

THE MOD-1 AND MOD-2 WIND TURBINES

On July 26, 1976, the Energy Research and Development Administration added the Mod-1 wind turbine project to the National Wind Energy Program. This giant wind turbine (Fig. 1–14) is rated at 2000 kilowatts at a wind speed of 25 mph (40 kmph) and has a two-bladed wind rotor with a diameter of 200 feet (61 meters). The power produced by this wind generator will increase with the wind speed from the start-up at a wind speed of 11 mph (18 kmph) to full rated power at a wind speed of 25 mph (40 kmph). As the wind speed increases above this speed the power output is held constant by varying the blade pitch. At a steady wind speed of over 35 mph (56 kmph) the machine is shut down.

NASA–Lewis Research Center was assigned by ERDA to manage the Mod-1 project, with General Electric's Space Division as the prime contractor. Other GE divisions will supply hardware as well as the 14,000-pound (16.350-kilogram) synchronous alternator. The Boeing Company will supply the two rotor blades, each of which weigh 18,000 pounds (8164 kilograms). The final design review of the machine was completed in August, 1977, and assembly began in March, 1978. The site for the first Mod-1 wind generator is near Boone, North Carolina.

The primary objective of the Mod-1 project is to develop a wind turbine of large enough size that the power produced will be competitive with the cost of electricity produced by other alternative energy sources. The final cost for the first Mod-1 wind turbine is expected to be about $3.3 million, or $1650 per kilowatt of wind generator capacity. The Mod-O wind turbine constructed at Plum Brook cost approximately $5000 for each kilowatt of generator capacity. On a yearly basis, the cost of power from the first Mod-1 wind turbine is expected to be 8.1¢ per kilowatt-hour.

The Mod-2 project is a further increase of wind generator size, with the goal of reducing the cost of power produced to the range of 2 to 4¢ per kilowatt-hour. The Mod-2 wind generator (Fig. 1–15) will have a rotor diameter of at least 300 feet (91 meters). The Boeing Company will develop the Mod-2 wind turbine, which is currently in the preliminary design stage. The machine will be designed for areas where the average wind speed is about 14 mph (22.5 kmph). A new type of tower is being developed, which will be a 187-foot (57-meter) steel shell braced with four tubular struts at the 50-foot (15 meter) level.

The Mod-1 and Mod-2 wind generators will utilize the most up-to-date materials and methods of design and fabrication. These machines will insure that the United States will be in the forefront of worldwide wind power utilization during the 1980's.

RECENT DEVELOPMENTS IN THE U.S. WITH SMALL WIND GENERATORS

The National Wind Energy Program has not neglected the development of small wind generators of the type that would be useful for supplying energy to a single home, ranch, or farm. In May, 1976, the Energy Research

Fig. 1–16: Rocky Flats, a national test site for small windmills. (Kaman Aerospace Corporation)

and Development Administration awarded a contract to Rockwell International for the development of a national test site for small windmills at Rocky Flats, near Golden, Colorado (Fig. 1–16). The first stage of this project was to establish a fully instrumented test site where Small Wind Energy Conversion Systems (SWECS) manufactured by private companies could be evaluated. To fulfill the major objective of the project, which is to stimulate the manufacture of small wind generators by private industry and to encourage their use by the public, personnel at Rocky Flats will install wind generators on towers provided at the test site, test and evaluate the machines, and develop an effective two-way communications link with manufacturers. New machines and modifications to existing wind generators can be tested and evaluated quickly under consistent conditions at the test site.

The test site is located on a large level plain that experiences a variety of wind conditions from gentle breezes to hurricane-force winds. The stronger winds predominate from one direction, and the towers were aligned in a pattern where the upwind machines would not block the airflow to other wind generators on the site (Fig. 1–17).

The first wind generator installed at the test site was a 4.1 kilowatt Aerowatt machine, originally operated by NASA at Plum Brook, Ohio. For the two and a half months it was tested, it generated power into a resistor bank, and powered a deep well irrigation pump, while exposed to winds as high as 85 mph (137 kmph). On February 23, 1977, a blade broke off near the hub, and the resulting imbalance fractured the casing of the machine, and the entire wind generator fell to the ground.

While failures of this type are always a disappointment, they help emphasize the value of a national test site with its open communications link between user and manufacturer. Eleven other small wind generators have been ordered or installed on towers for testing at the site, including a rebuilt Jacobs 3 kilowatt machine. No wind generator tested at the heavily instrumented site has given entirely faultless operation, and problems have ranged from broken tail vanes and defective bearings to the loss of blades. But Rocky Flats is having a positive effect on improving wind generators marketed by private manufacturers, as several companies are

Fig. 1–17: Site plan for Rocky Flats experimental wind generator installations.

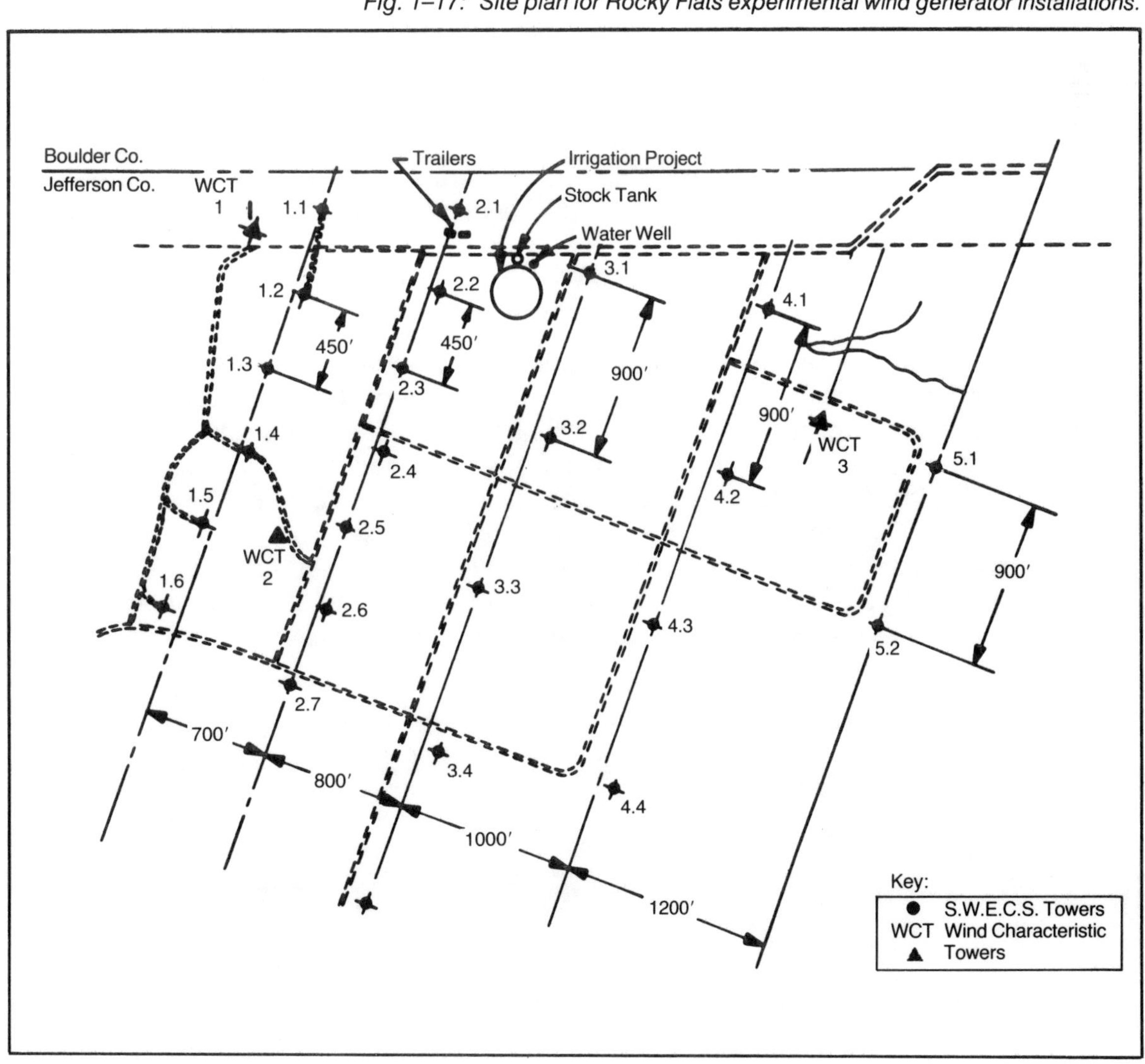

redesigning their machines to incorporate recommendations made by the technical experts at the test site.

Technology development is another part of the program at Rocky Flats. Outside organizations will be funded to design and develop small wind generators in three distinct power ranges—1 kilowatt, 8 kilowatts, and 40 kilowatts. These machines will satisfy a number of needs which have been clearly identified: an 8 kilowatt general purpose alternating current wind generator for individual home and farm use; a 40 kilowatt alternating current machine for powering irrigation pumps and to provide electricity for isolated small communities; and a small 1 kilowatt direct current machine with battery storage for very remote and severe climate locations where the cost of a conventional generating station would be prohibitive. Some applications in this last category include communications repeater stations, remote weather and earthquake monitoring stations, and coastal and offshore navigation beacons. Contracts have been awarded to private companies to develop these machines, and they will be tested on towers at Rocky Flats when the manufacturers complete construction of the first models.

RECENT DEVELOPMENTS IN OTHER COUNTRIES

Since the mid-1970's other countries besides the United States have begun to look towards wind energy conversion as a practical source of electricity. Some countries, such as Denmark and the Netherlands, once used wind power extensively but abandoned this source of energy when oil prices plummeted after World War II. Other countries, like Canada, are exploring wind power on a national scale for the first time.

As oil and natural gas production throughout the world peaks and then begins to decline, energy prices will rise, and other countries will also turn to wind power for at least part of their energy needs. This situation surfaced in Canada during 1975, when Canada passed from a net exporter to a net importer of oil to satisfy the increasing domestic demand for energy. Part of Canada's answer to the energy crisis came in May, 1977, when the National Research Council and Hydro Quebec began the final assembly of the first large-scale wind generator in Canada, the 230 kilowatt Magdalen Islands wind turbine, located in the windy Gulf of St. Lawrence.

The Magdalen Islands wind turbine does not resemble the familiar two-, three-, or four-bladed windmills pictured previously in this chapter. It is a modern adaptation of the first windmills, the vertical-axis machines. The Magdalen Islands generator (Fig. 1–18) is based on a vertical-axis wind turbine with curved airfoil blades invented by G. J. M. Darrieus in France during the 1920's.

Fig. 1–18: Darrieus vertical-axis wind turbine.

Modular adaptation of Darrieus design.

ERDA wind driven generator at Clayton, New Mexico. (Department of Energy—NASA)

Chapter Two

Wind

Wind is air set in motion by the uneven heating of the earth by the sun. The earth and its envelope of air, the atmosphere, receives more solar heat near the equator than at the polar regions. Yet, the equatorial regions do not become hotter each year, nor do the poles become colder. It is the movement of air around the earth that averages temperature extremes and produces the surface winds so useful for power generation.

In its vertical dimension, the atmosphere is divided into a number of layers. The layer adjacent to the earth is called the *troposphere.* It ranges in height from about 60,000 feet (18,300 meters) over the equator to about 25,000 feet (7,600 meters) over the poles, with greater height in summer than in winter.

Antartic installation—a Dunlite 2 KW brushless wind-driven generator

Although extremely light, air has weight and is highly elastic and compressible, like all gases. A cubic foot (0.28 cubic meters) of air at ordinary temperature and pressure weighs 1.22 ounces (35 grams). Pure, dry air contains about 78 percent nitrogen, 21 percent oxygen, and a 1 percent mixture of 10 other gases. The proportions are about the same in all parts of the world.

Like all gases, air expands, or increases in volume, when heated and contracts, or decreases in volume, when cooled. In the atmosphere, warm air is lighter and less dense than cold air and will rise to high altitudes when strongly heated by the sun. Warmed air near the equator will flow upward and then outward towards the poles where the air near the surface is cooler. The regions of the earth near the poles now have more air pressing down on them, and the cooler surface air tends to slide away from these areas and move towards the equator.

THE CIRCULATION OF THE ATMOSPHERE

The movement, or circulation, of the atmosphere which actually results from uneven heating is profoundly influenced by the effects of the rotation of the earth, which is about 1000 miles (1600 kilometers) per hour at the equator, decreasing to zero at the poles; and by the proportion of land to sea area, location of the continents, and such features as mountain ranges. In general, surface winds do not blow directly and steadily from colder to warmer regions, and upper winds from warmer to colder regions. And, as is to be expected, the periodic variations in the distribution of solar heat with the seasons of the year gives rise to similar variations in the circulation.

The actual wind pattern that results is shown in Fig. 2–1. Both north and south of the equator the circulation breaks into three distinct zones, called *cells.* The first is the tropical cell which extends from the equator to about 30° north and south latitude. The air over the equatorial region is heated by the sun and rises to the top of the troposphere, the *tropopause,* and starts moving towards the poles. At the surface the equatorial region is called the *doldrums.* The air is hot and sultry, and any winds quickly sink to stagnation, but light breezes may develop from any direction. The sky is overcast and showers and thunderstorms are frequent. Air pressure remains consistently low.

The position and width of the doldrum belt varies with the season. During February and March the doldrum belt hugs the equator and is very narrow. By July and August the doldrum belt may have moved farther north, and may become quite wide, even covering several degrees of latitude at its narrowest point.

At the northern and southern fringe of the tropical cells, at about 30° north and south latitude, there is the subtropical high pressure belt of subsiding or sinking air that descends to the surface creating a zone of dry, warm air called the *horse latitudes.* Winds are light and variable, and may be stagnant for several days. Almost all of the earth's deserts are located in these regions. The name horse latitude dates to when the

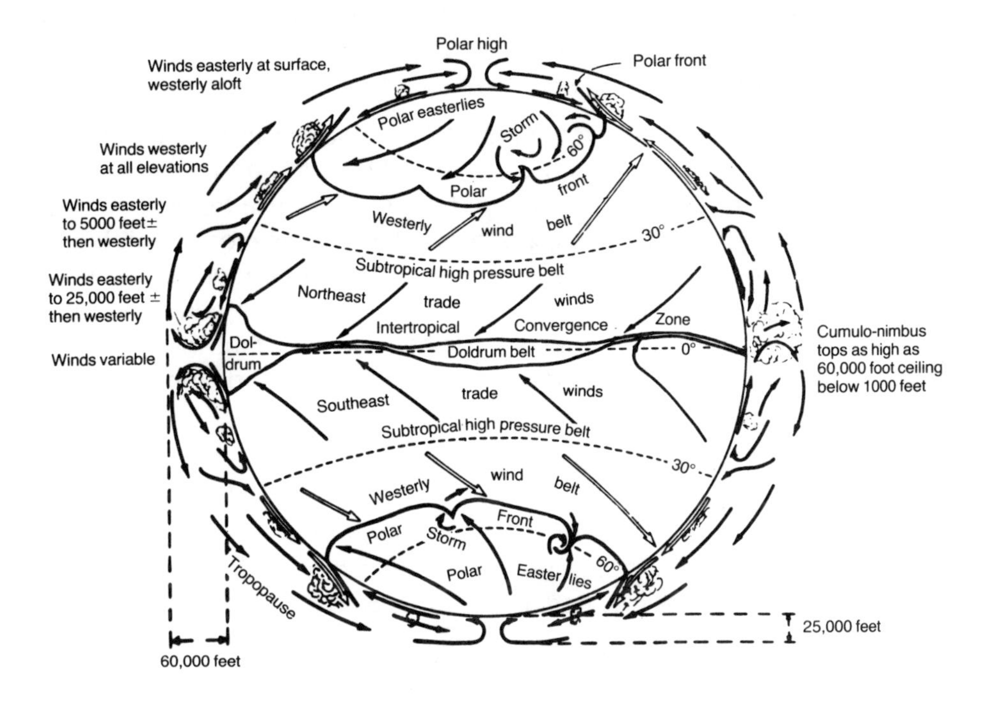

Fig. 2–1: World wind circulation

Western Hemisphere was being colonized. The crews of sailing ships, becalmed for long periods in the light and variable winds, were forced to throw horses overboard to preserve their drinking water supply.

Sinking air over the horse latitudes spreads out in both southerly and northerly directions. In the Northern Hemisphere, the southerly moving winds become deflected to the west and become the persistent northeast trade winds. (Winds are identified by the direction from which the wind is coming.) The trade winds are the most constant of winds, for they often blow for days or even weeks with only slight variation in direction and strength.

The northerly winds from the horse latitudes become the westerly wind belt between 30° and 60° north latitude. Only in the Southern Hemisphere do these westerly winds exhibit a persistence approaching that of the trade winds. Their course in the Northern Hemisphere is interrupted by the greater land areas and mountain ranges of this area compared to the large expanse of ocean in the Southern Hemisphere. Fig. 2–2 illustrates prevailing ocean winds during January and February recorded from shipboard stations. Note how the westerly wind belt at 40° south latitude is undistorted by land areas compared to the westerly wind belt at 40° north latitude.

The polar cells extend from about 60° latitude to each pole. Each polar cell is the result of the extreme coldness in these regions, particularly during the long winter night when there is no solar heating. These regions of cold, dense air produce surface winds called the polar easterlies which spread toward the equator from the fringe of the polar cell, the polar front. These easterly winds are literally pushed out of the polar region by the high pressure near the poles as high altitude westerly winds cool and subside over the poles. It is the warmth brought by these high altitude winds that keeps the polar region from becoming colder each year.

The most violent and varied weather takes place in the temperate, or mid-latitude, mixing cell between 30° and 60° latitude, particularly in the Northern Hemisphere. Over this region tongues of warm tropical air and cold polar air meet, producing a climate over land areas most favorable to agriculture. Since human progress as we know it depended on the successful development of agriculture where only a small percentage of the population could provide food for all, it is no accident that the most advanced civilizations have flourished beneath the mid-latitude mixing cells.

WIND OVER LAND AREAS

Worldwide wind patterns are profoundly altered over land areas. Each major land area has its unique wind patterns which result from factors that control the general climatic pattern of the area. Some of these factors are:

1. Latitude; shape and size of the land area
2. Coastal and inland water distribution
3. Topography of the land area

Because land and water heat and cool at different rates, the location, shape, and size of land areas greatly alter the earth's wind patterns. The upper layers of the oceans are nearly always in a state of violent stirring, thus heat losses or heat gains occurring at the ocean surface are distributed throughout large volumes of water. This mixing process sharply reduces the temperature contrasts at the surface between day and night, and between winter and summer.

Fig. 2–2: Prevailing ocean winds.

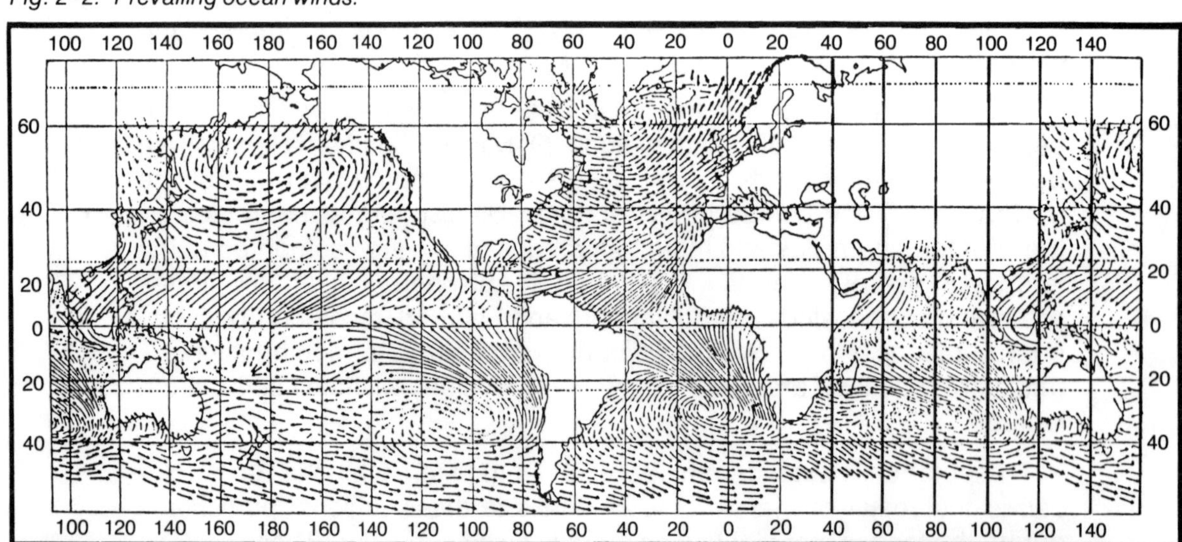

Key: Arrows fly with the wind
Length denotes measure of steadiness
The darker the arrow, the greater the force

On land, however, very little heat penetrates into the soil, and then only to a shallow depth. There is no distribution of heat through the soil by mixing, as with water. Soils warm during the day and cool quickly at night. Air above soil will also quickly warm and cool. Thus, strong contrasts are imparted to the air above land areas between day and night, and between the seasons of the year. The daily and annual spread of temperature and wind is much larger over land than over water.

Coastal areas take on the climatic characteristics of the land or water in the direction of the prevailing winds. In the latitudes of the prevailing westerly winds, for example, the west coasts of land areas have the uniform temperature and wind patterns of the adjacent ocean. East coasts are more strongly influenced by wind patterns generated over the interior of the land area.

The topography, or physical features of the land area, can strongly influence the wind patterns. Mountains hinder uniform wind patterns—air channeled around or through gaps often gives rise to strong local winds ideal for wind power generators. With long mountain ranges perpendicular to the westerly wind belt, such as the Rockies and the Andes, the wind cannot go around and must go over, producing unusual wind patterns on the lee side of the mountain ranges.

LOCAL WINDS

Many of the winds that are encountered in day to day living are of strictly local origin and are almost entirely due to local differences in temperature and the influence of local topography.

Along coastal regions under clear daytime skies, the temperature of the land will rise while the temperature of the water will remain relatively constant. The warm land heats the air above it, and as the air above the land becomes less dense it rises. The cooler air over the water moves over the land to replace the rising warm air, and a circulation is established (Fig. 2–3). This circulation produces the cool onshore breeze that makes seaside areas comfortable during the warm summer months. At night the situation is reversed. The land cools rapidly, and if the air over the water is warmer than that over land an opposite circulation is estab-

Fig. 2–3: Land-sea breeze

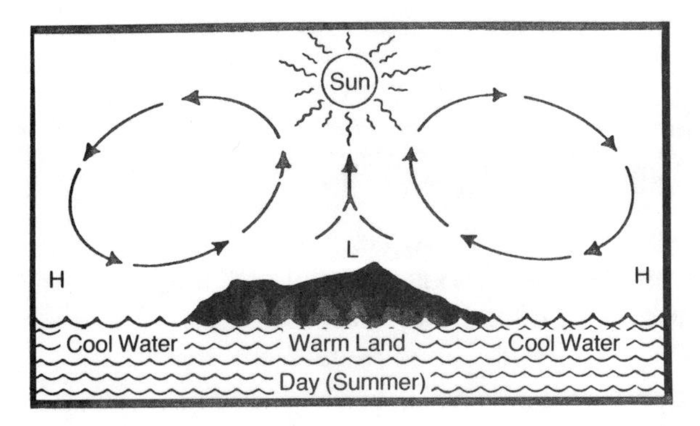

Fig. 2–4: Mountain-valley breeze

lished, this time with the breeze blowing from the land out over the water. This is called an offshore breeze.

Another type of local wind is the mountain breeze. During daylight hours, when there is a strong sun and no prevailing wind pattern, the valleys and the sides of the mountains become warmed by the sun. As the air above this ground warms it rises up the mountain slopes while denser air in the center of the valley settles. This denser air in turn is warmed as it settles and rises up the slopes, and a circulation develops (Fig. 2–4). This is a valley breeze, called an *anabatic wind* (ana = up; bata = to go). At night the situation is reversed. The sides of the valley cool quickly, chilling the air immediately above the surface. This air is more dense than the air farther from the surface, and flows downhill to the bottom of the valley, creating a gravity wind, since the denser air is being accelerated down the slope by the earth's gravity.

Local winds that blow down a slope are called *katabatic winds* (kata = down; bata = to go). This term includes any wind blowing down a slope where the slope has an active part in generating the wind. Examples of katabatic winds include the *Chinook, foehn, bora, mistral,* and the *Santa Ana.*

The Chinook (North America) and the foehn (Europe) are local names given to large scale katabatic winds that race down the lee side of mountain ranges. As the air rises over the mountains it cools and the moisture it carries condenses and falls out as precipitation. The air reaching the bottom of the leeward side is much drier and warmer than the original ascending air and often reaches a high wind speed on the downslopes (Fig. 2–5). The warming effect of Chinook winds in winter is often felt several hundred miles from the crest of the mountains.

The bora and mistral are local names for katabatic winds common to the northern coast of the Mediterranean. These winds result from an extremely cold dense air mass at the top of a plateau that spills over and descends down the adjacent slopes. The Santa Ana winds result from dry air in the elevated lands of Nevada and northern Arizona that descends to

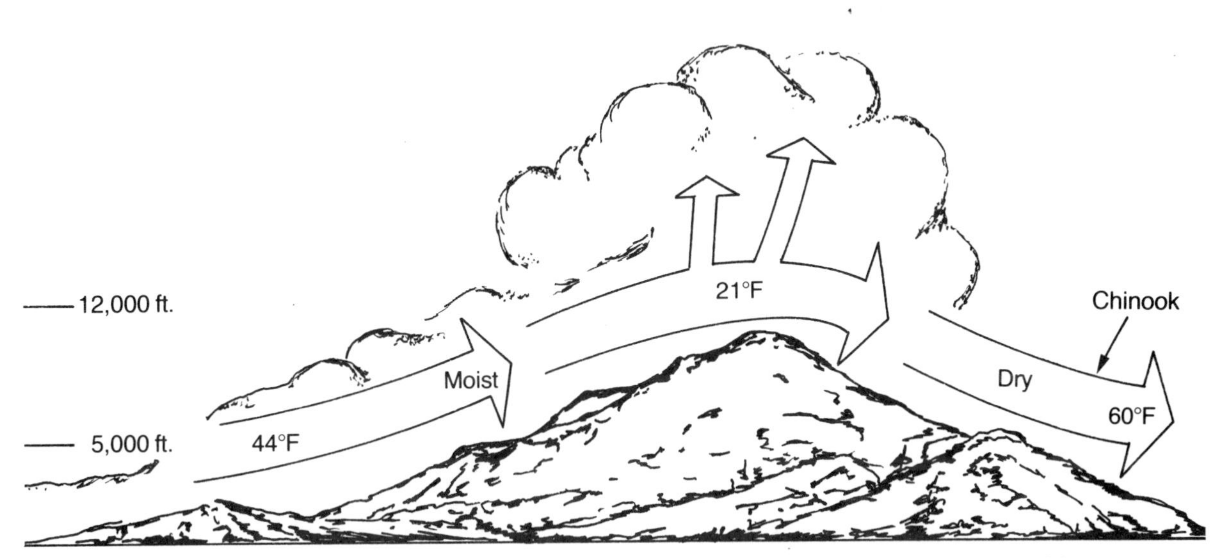

Fig. 2–5: Formation of Chinook winds

the coast of southern California. The air grows warmer and drier as it descends, and the strong wind velocities common to katabatic winds greatly increase the fire hazard in this dry area.

AIR MASS WINDS

In addition to the circulation wind patterns of the earth and local winds, winds useful for power generation result from the movement of air masses. Air mass movements occur primarily in the mid-latitude mixing cell, and the source of all important air masses is either the polar cell or the tropical cell.

An air mass is a huge body of air usually 1000 miles (1600 kilometers) or more across of uniform temperature and humidity. Warm air masses invade the mid-latitude mixing cell from the tropical cell and predominate in summer. During the winter cold air masses from the polar cell predominate.

Air mass movements into the mid-latitude mixing cell are predictable seasonally but cannot be forecast on a day-to-day basis. In North America, for example, air mass movements (Fig. 2–6) are responsible for changes in the prevailing wind direction recorded at Midwestern weather stations between winter and summer. Winds from the north and northwest predominate in winter, while winds from the south and southwest predominate in summer.

The study of air mass movements provides the basis for weather forecasting in most land areas under the mid-latitude mixing cell.

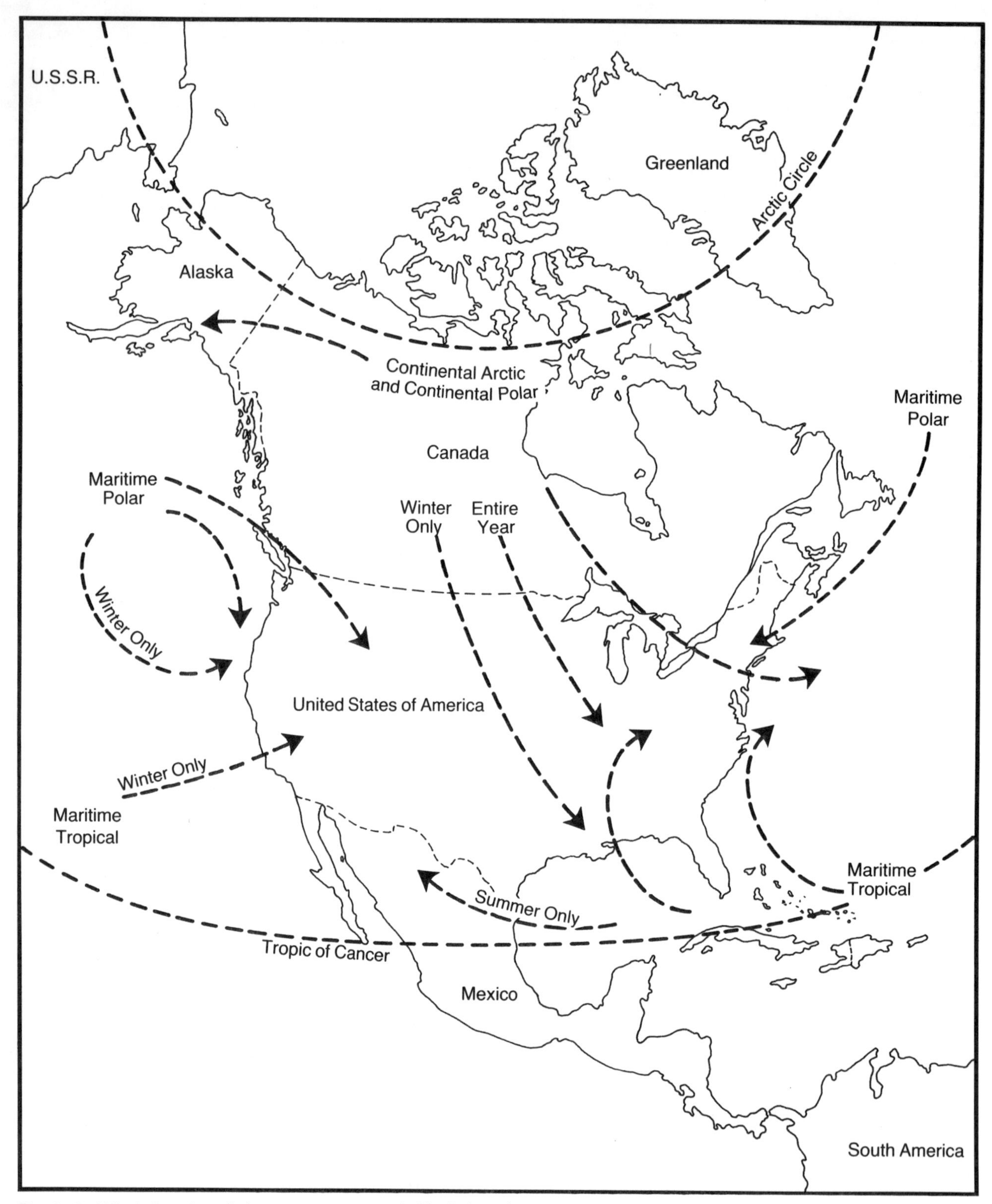

Fig. 2–6: Wind direction of air mass movements in North America

Monitoring ambient conditions for industrial application. Wind speed/direction, temperature, dewpoint, pressure, and solar radiation. (Climet Instruments)

WINDS OF SELECTED LAND AREAS

In this section typical wind patterns of large land areas will be discussed. Where wind patterns at specific seasons of the year are mentioned it should be noted that these wind patterns do not necessarily persist during the entire year.

Africa

Africa is unique among continents in that it extends almost equally north and south of the equator, to approximately 35° north and south latitude. Africa lacks the extensive mountain ranges in locations where they act as climatic barriers that other continents feature, but Africa does have vast areas of plateau at elevated altitudes.

The winds of North Africa are influenced by wind patterns over the Mediterranean and tend to be primarily westerly (Fig. 2–7). Three wind patterns from the north and northeast emerge over much of Africa be-

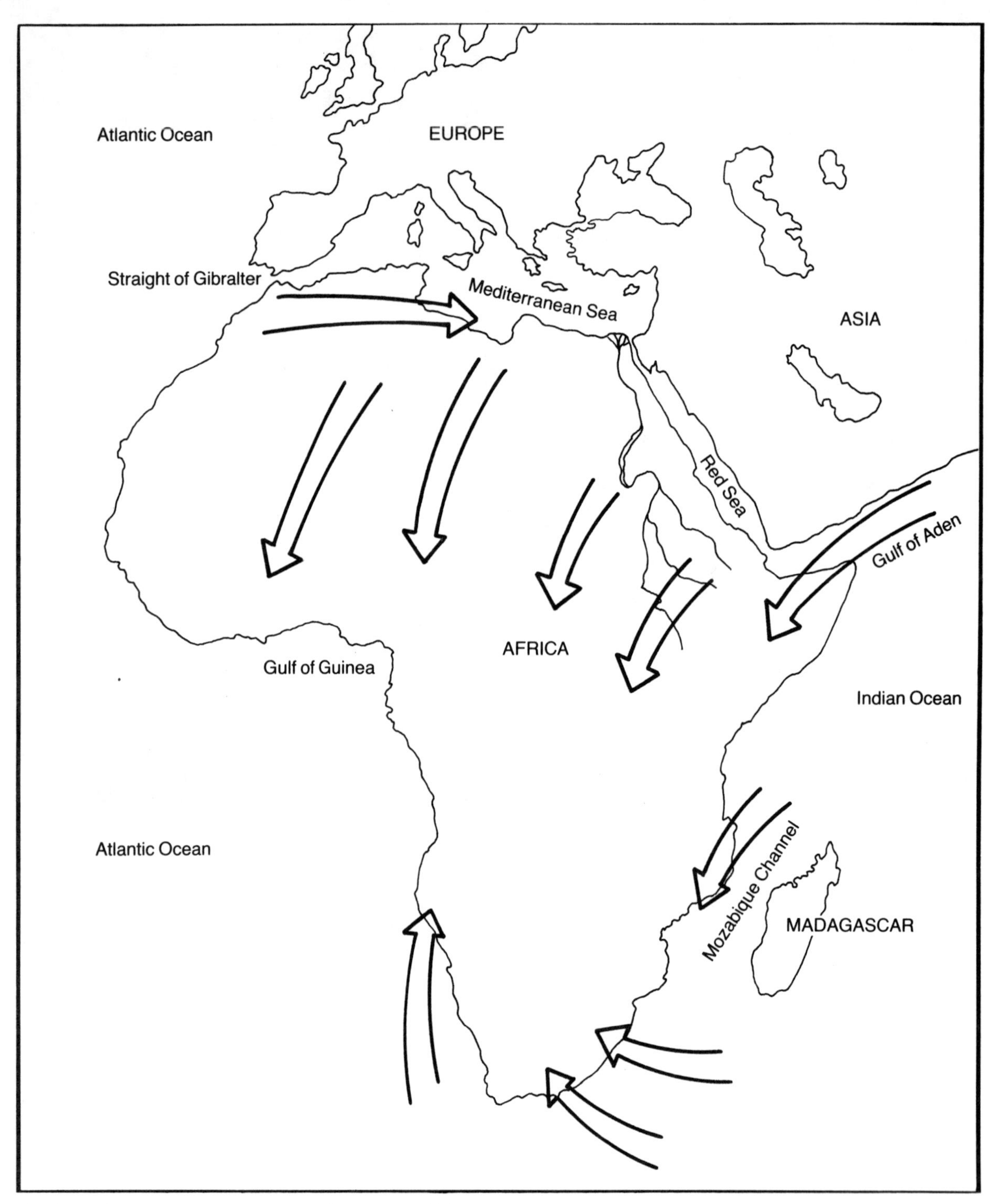

Fig. 2–7: African wind patterns (December–February)

tween November and March. A northeast trade wind develops over the Atlantic coastal area at the bulge of Africa, northerly winds press down over the interior, and a northeast wind that originates over Asia and the Pacific spreads across the eastern coastal region. South Africa is under the influence of different winds; the southeast trades of the Indian Ocean on the east coast, and southeast trade winds to the west which become deflected along the coast, turning southwesterly inland.

The inland winds from the north and northeast across the northern portions of Africa are the *harmattan,* the continental trade winds that blow as a moderate to strong wind during the day, but drop to calm during the night.

From May to September the harmattan recedes to the north, and southwest and southeast winds from the surrounding oceans blow far inland. April and October are transition months, and winds tend to be light and variable in many parts of Africa. The topography of Africa affects local wind patterns, particularly in the plateau of East Africa. At an altitude of over 5,000 feet (1,525 meters) the plateau is above surface winds from the ocean and is affected more by upper wind patterns. In general, Africa is a continent of predictable winds throughout the year, steady during daylight hours but dropping off to near calm at night.

Asia

Asia, the largest continent, is situated in the middle latitudes of the Northern Hemisphere, with only the peninsulas of India and Southeast Asia projecting toward, but not reaching, the equator. As would be expected, the wind patterns of these peninsulas are strongly affected by the adjacent ocean area.

India is primarily a region of light to moderate winds, due in part to the Himalayas to the north which act as a climatic barrier, protecting India from strong continental winds from the north. The winds of the low level plains of India are primarily northeast trades, with steady onshore winds during much of the year (Fig. 2–8).

The dominant winds of the peninsula of Southeast Asia originate as the north Pacific trade winds during winter in the Northern Hemisphere, while during the summer season the south Pacific trade winds predominate. But, since this region lacks the effective climatic barrier of the Himalayas which protect India, north winds from Siberia reach into Indochina, particularly during the winter. In the islands south of Singapore hot winds from south of the equator are frequent, but local land and sea breezes and mountain and valley winds mask general wind patterns.

Cold polar winds pour over the interior of Asia, from the north and northwest in north China, while north and northeast winds affect the south China coast. To the west, the prevailing westerly winds of the Mediterranean are prominent, while north and northwesterly continental winds blow steadily over much of the inland regions.

Europe and England

Much of the topography of Europe is composed of projections of land surrounded by sea: Norway, Sweden, and Denmark in the north, Spain to the southwest, Italy and Greece to the south, and, of course, the British

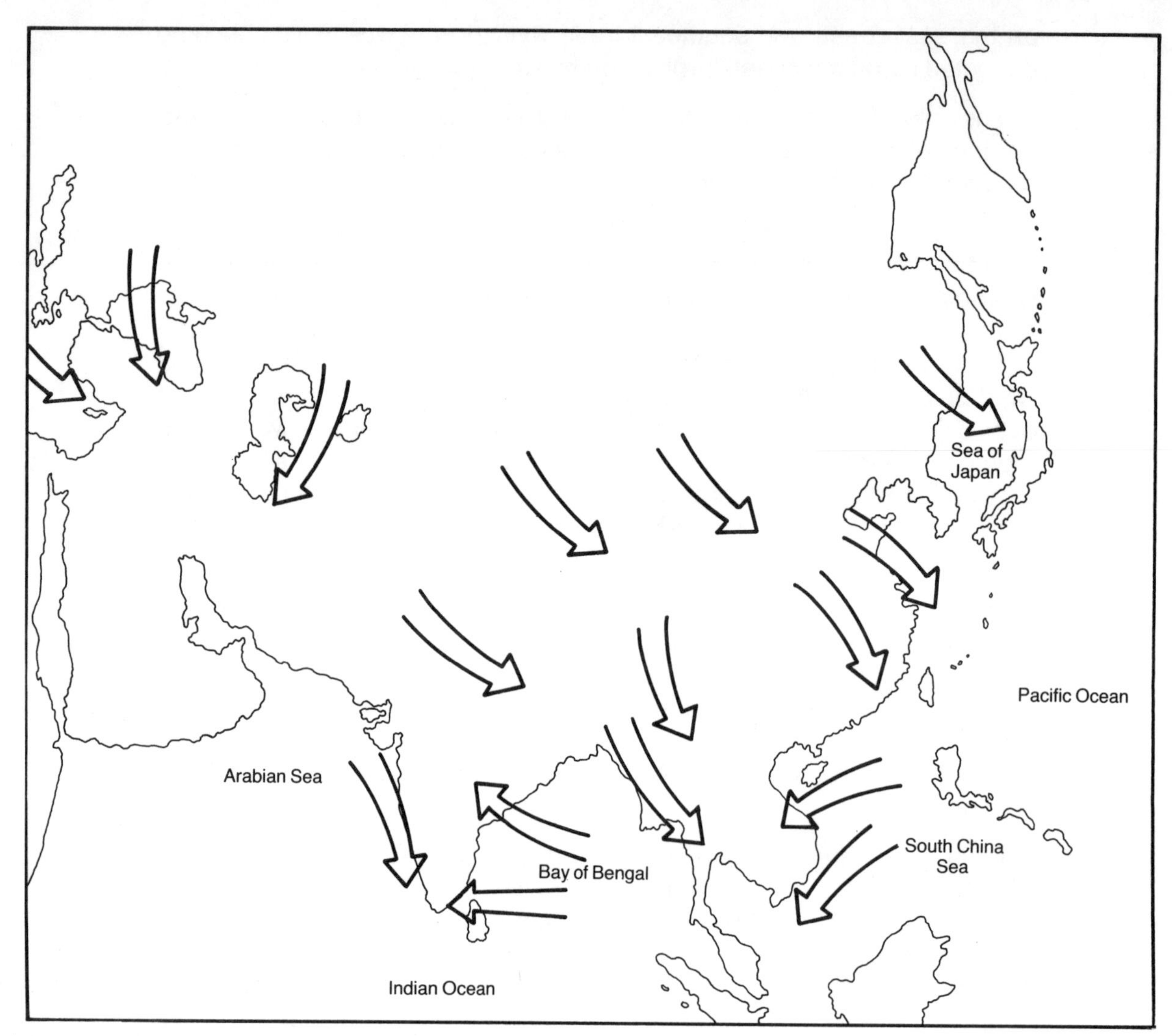

Fig. 2–8: Prominent winds of Asia

Isles. Even France may be thought of as a peninsula, with the Atlantic to the north and the Mediterranean to the south. Thus, local land and sea breezes are much in evidence, and the absence of mountain ranges along the west coast, as in North America, ensure an unimpeded path to ocean winds which are felt far inland.

The dominant winds (Fig. 2–9) are the seaborne westerlies, but since most of Europe falls within the mid-latitude mixing cell, outbreaks of cold polar air and warm tropical air spell variable winds from practically any direction, including the east. In general, the prevailing winds tend to blow from the southwest in winter and from the west and northwest in summer. The southwesterly winds of winter pass over warm ocean water and give more warmth inland than the weak rays of the sun. Winds are stronger in winter than in summer, when local winds such as land-sea breezes mask the dominant wind pattern.

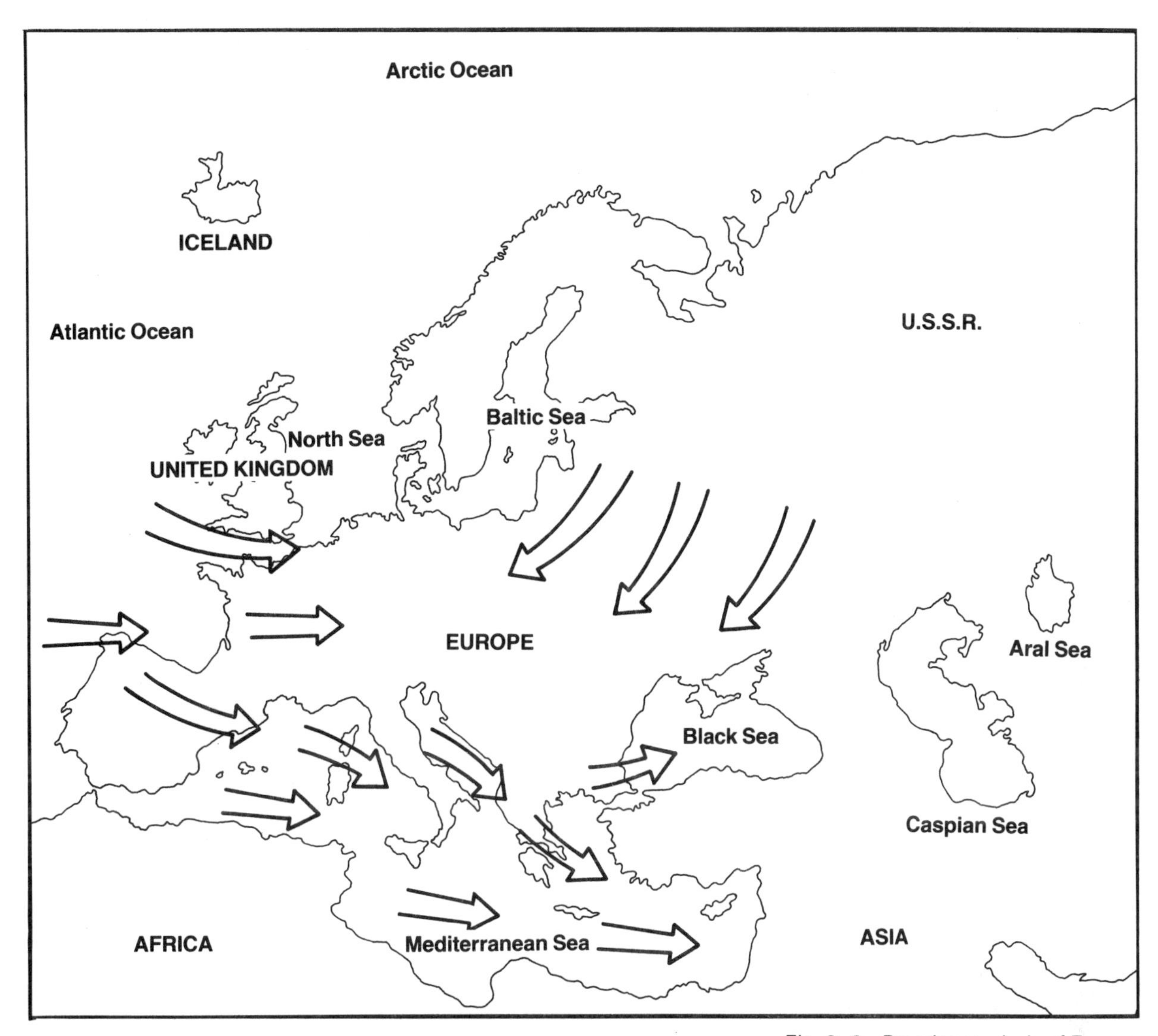

Fig. 2–9: Prominent winds of Europe

The westerly winds are neither as constant nor as strong in central Europe, which is the transition region between oceanic westerlies and the cold north winds from Russia. The Alps lie in a wide east-west path from 0the Mediterranean coast of France to near Vienna, Austria, and block the cold polar winds of central Europe from invading the lands to the south.

Westerly and northwesterly winds dominate much of the Mediterranean region, but local land-sea breezes greatly modify the general direction. Winds are moderate to strong over much of this region, especially in the region of Greece and the Greek islands, where more northerly winds blow with such force that in many places trees cannot grow on high ground. In addition, strong local north winds are felt over other areas of the Mediterranean in winter, replaced with strong southerly winds from Africa during the spring.

Australia

Australia is the smallest of continents, and the driest. Over a third of Australia has less than 10 inches (25 centimeters) of rainfall annually, while only about nine percent has more than 40 inches (100 centimeters). Although the terrain of Australia does not favor rainfall, the higher plateaus in the east and west receive more rainfall than the lowlands that split the continent from north to south.

Much of Australia is under the influence of the southeast trade winds, which are relatively dry winds and help account for the dryness of the continent (Fig. 2–10). The windward slopes of the Eastern Highlands receive the most rainfall, while the regions to the west are the driest. The winds of Australia are most unsettled along the south coast and the south of western Australia, where the dominant westerly and southwesterly winds are interrupted by cold polar winds from the south.

Land-sea breezes account for the coastal areas being the windiest, and mountain-valley winds are particularly evident along the eastern and southeastern coasts, from about 22° south latitude to Melbourne, where cool katabatic winds from the highlands sweep over the coast. In northern Australia during summer in the Southern Hemisphere the southeast trade winds weaken in strength, and northwesterly winds become dominant. This is the rainy season in the North. By April and May the southeast trade winds are again dominant. Along the coastal regions winds from the sea are strongest during daylight hours, dropping to calm at night.

Australia is subject to unusually strong local winds not found in other regions of the Southern Hemisphere. These include the tropical cyclones of the northwest, the *Willy-Willies,* that originate over the warm sea north of Australia and travel first southwest and then southeast far into the continent. Similar tropical cyclones originate near the Fiji Islands and travel southwesterly to the eastern coast of Australia. Hot winds from the deserts of the interior travel to the south coast and may last for many days, bringing temperatures of over 100° F. to normally cool coastal towns. These winds are called locally *Brick Fielders.* A similar phenomenon, but with cold air, is called the *Southerly Bursters,* and results from an outburst of polar air from the south replacing the hot continental air from the north. Bursters occur in the spring and summer, with winds occasionally reaching gale force.

South America

South America shares the Southern Hemisphere with Africa and Australia, but has a distinctive climate of its own. The Andes, the north-south mountain range, split most of the continent into western and eastern wind patterns, and create strong local wind patterns. South America tapers toward the polar region, and thus is less affected by polar outbreaks than continents in the Northern Hemisphere with large polar land areas.

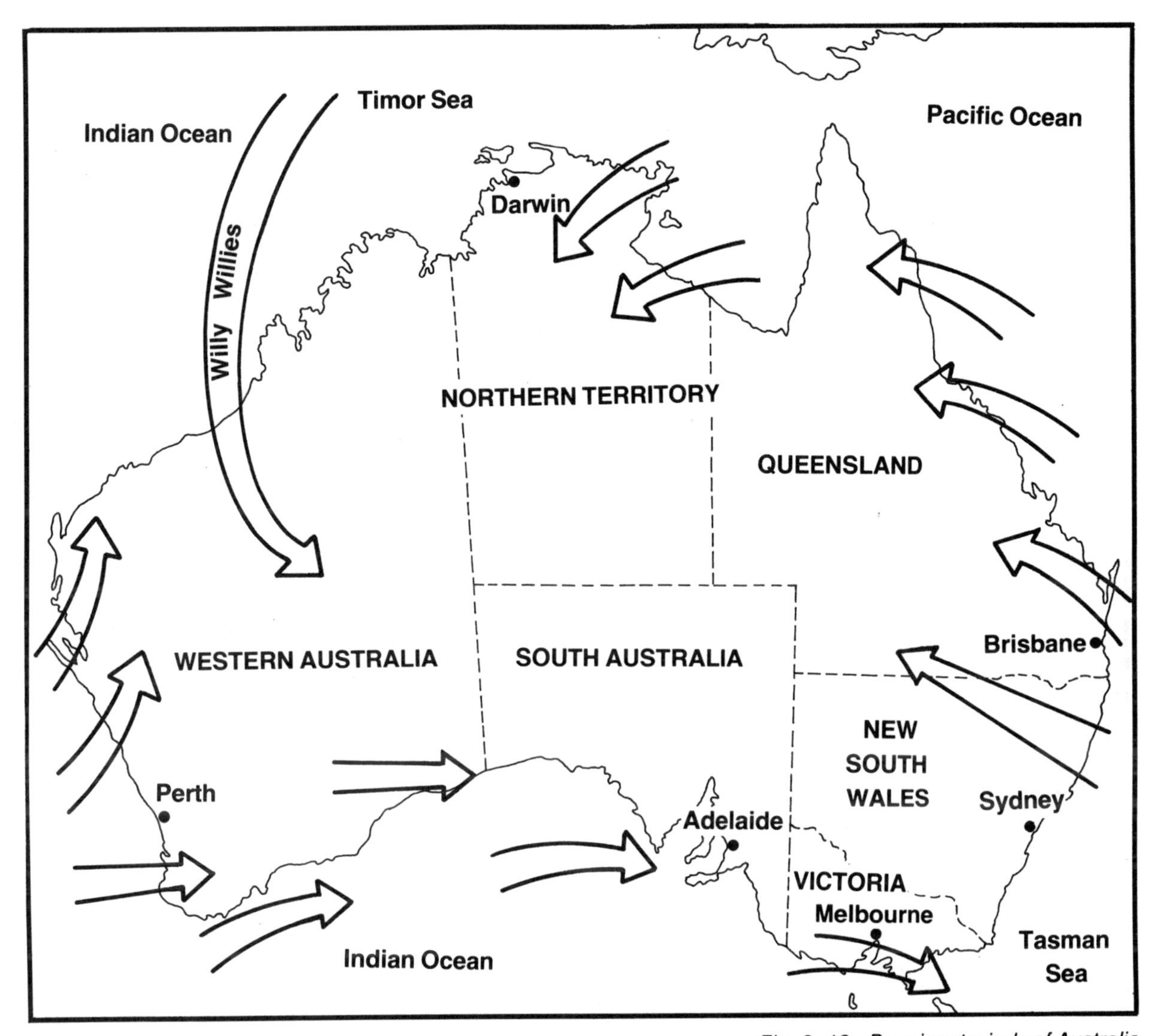

Fig. 2–10: Prominent winds of Australia

The narrowness of the continent to the south limits temperature extremes, and most winds are of maritime origin. These winds (Fig. 2–11) include the northeast trades in the region north of the equator, and the southeast trades farther south on the east coast. These winds travel far inland, particularly in the Amazon basin of Brazil.

Beyond 38° south latitude the prevailing westerlies cross the continent without great difficulty, as the Andes are much lower here than farther north. The southwestern tip of the continent is among the windiest places on earth, with average wind speeds of between 25 and 35 miles (40 and 56 kilometers) per hour in many locations.

Much of the west coast of South America is dominated by the Pacific trade winds, which blow in from the southwest north of the westerly wind belt and are deflected northward along the coast. The Andes effectively block all Atlantic and interior winds from the west coast region. Over the

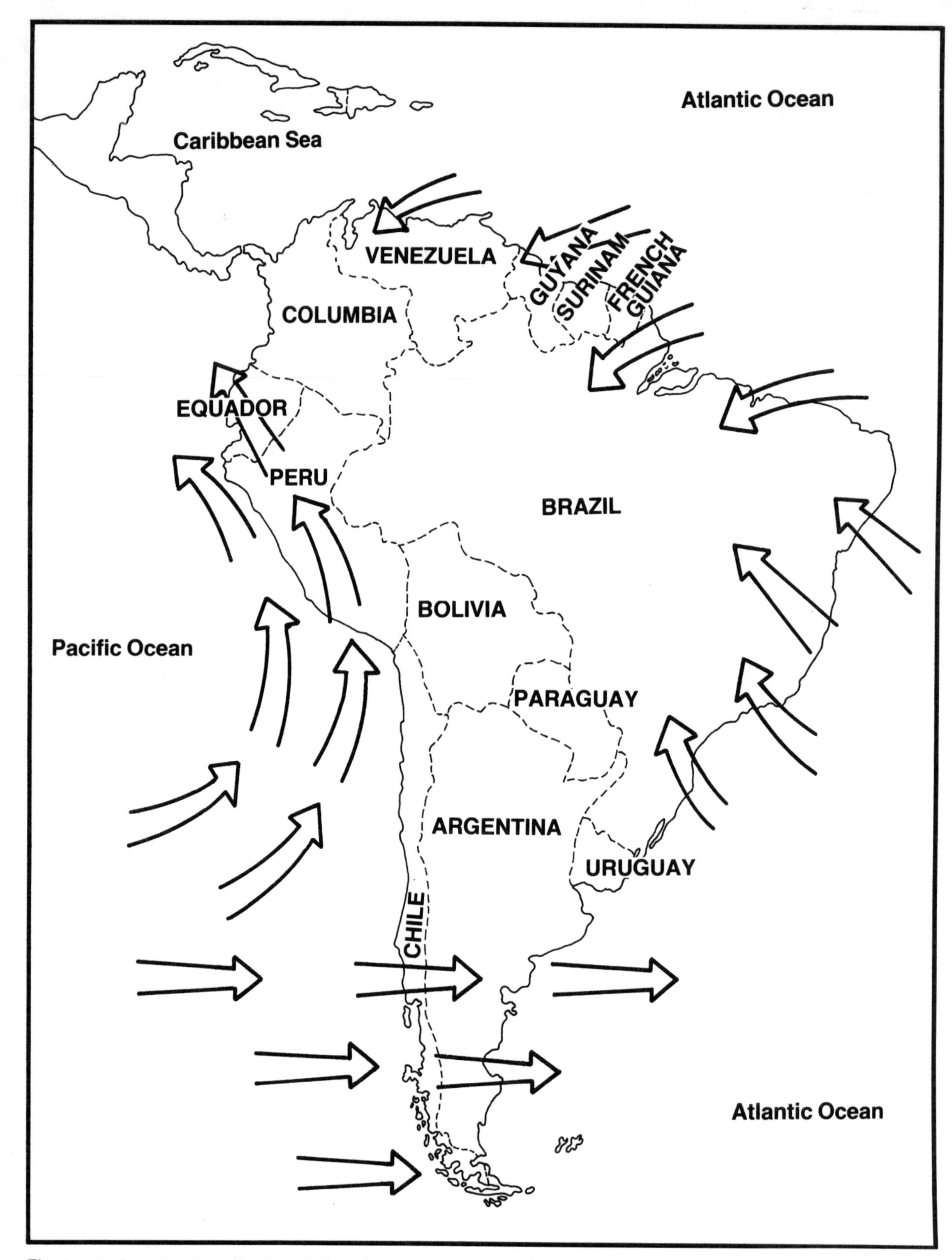

Fig. 2–11: Prominent winds of South America

west coast in particular land-sea breezes are very strong, the sea breeze during the afternoon so strong as to stop near–shore boating at times. Also present is a variation of Chinook winds, due to the lofty mountains near the coast.

North America

The outstanding feature of North America is its north-south mountain range near the west coast which blocks the travel of westerly winds into the interior. As much of North America is within the mid-latitude mixing cell, and the Rocky Mountains block strong westerly wind movements from the Pacific, the stage is set for the interior to serve as a vast staging ground for cold polar outbreaks from the north and warm humid air from the south. Unlike land areas influenced by prevailing sea winds, wind directions change with cold and warm air masses, and with the seasons of the year (Fig. 2–12). In the Midwest, for example, the prominent wind direction is from the northwest during winter and from the southwest during the summer.

Due to the lack of prominent east-west mountain ranges (the Appalachians are too low) wind patterns from alternating cold and warm air masses spread throughout the interior. In winter, it is not uncommon for the winds of cold air outbreaks to spread from Hudson Bay to the Gulf of Mexico.

Local winds include Chinook winds on the lee side of the Rocky Mountains, which may be felt for several hundred miles (up to 500 kilometers) to the east, and land-sea breezes along the coastal regions. The Great Lakes do not have a major effect on wind patterns. Their primary influence is to moderate winter temperatures. Hudson Bay is largely frozen and has much less influence on the wind patterns of surrounding land areas than interior seas of other continents. Among the windiest areas are the St. Lawrence River region and Newfoundland, as many continental air masses pass out to sea through this region. This area is also affected by strong northeast and east winds from the cold North Atlantic Ocean.

WIND POWER SITES

The preceding introduction to the nature of world-wide, continental, and local wind patterns is a fundamental first step in considering the location of wind power sites. On a continental scale, the coastal regions of land areas and locations that fall within the belt of strong westerly winds or trade winds, such as the southern part of South America, offer the greatest potential as wind power sites (Fig. 2–13). On a local scale, knowledge of the prominent wind patterns is also important, as areas with prevailing winds from one direction usually have greater power generating potential than areas where the winds are variable in direction. In addition, a knowledge of prevailing wind patterns will aid in identifying natural and manmade obstacles in the path of wind power sites, such as changes in terrain, forests, and large buildings.

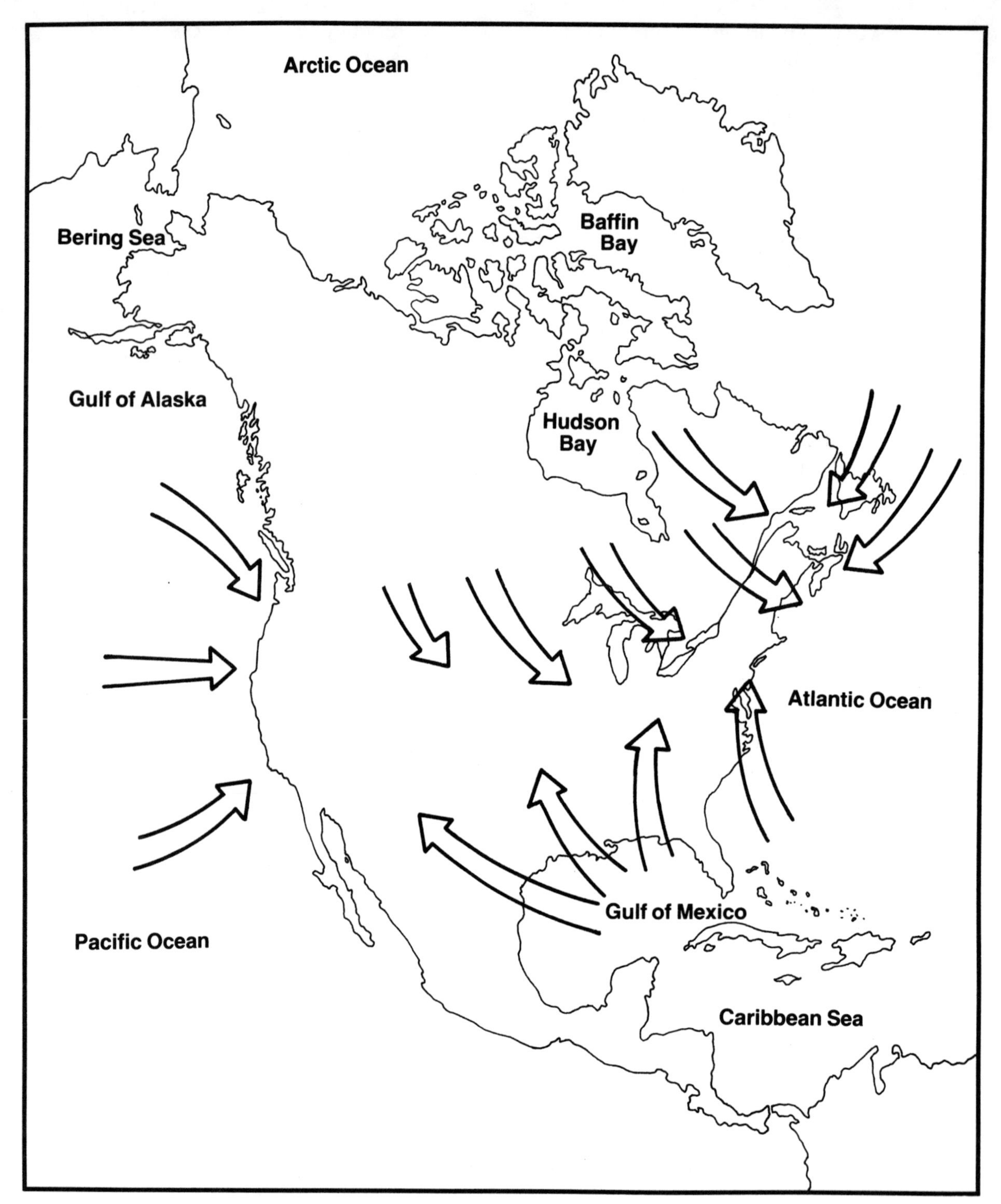

Fig. 2–12: Prominent winds of North America

Importance of Site Selection

There is evidence that crude windmills were used in China more than 2000 years ago, but only in the past 50 years has the importance of choosing the windiest available site been recognized. The power available from wind increases greatly as the wind speed increases, and it is the goal of site selection to find a location with the greatest average wind speed. A site with a wind speed of even one or two miles per hour higher than the surrounding area means a significant increase in the annual output of electrical energy. For example, there is 338 percent more power available from wind with a speed of 12 mph (19 kmph) than from wind of 8 mph (13 kmph).

Selecting the best site for a windmill is done in several ways. The meteorological records of the country can be studied to determine the most promising areas, then these areas can be evaluated with on-site measurements. In many areas certain types of terrain are more windy than others, and a knowledge of how wind behaves over different types of terrain is essential to evaluating potential wind power sites.

Meteorological Records

In many countries meteorological records which include wind speed and direction are available from a central receiving point. In the United States, this is the National Climatic Center (Federal Building, Asheville, NC 28801). The National Climatic Center collects, processes, and stores weather information from approximately 2000 locations. Data is collected from the National Weather Service, military services including the Coast Guard, the Federal Aviation Administration, the merchant marine, and privately owned airfields. The primary network consists of about 700 stations which record observations 24 hours each day. The remaining locations take observations during daylight hours only; every 3 or 6 hours; or to meet specific local needs.

Stations that have supplied wind summaries based on five or more years of data to the National Climatic Center are shown in Fig. 2–14. The wind summaries from these stations are most accurate for those locations east of the Rocky Mountains. This is because most of the stations are at airport locations. In hilly areas most of the stations, and of course, the airports, are located in level valley sites normally much less windy than nearby hilltops.

From its beginning, the National Weather Service (the Signal Corps of the U.S. Army prior to creation of the U.S. Weather Bureau) has utilized the rotating cup anemometer for recording wind speed (see Source/Resource section). Although minor changes in design and recording methods have been made over the years, this measuring instrument has remained basically the same. The greatest problem in interpreting wind data is, however, knowing the height above the ground of the anemometer. Although an attempt to standardize anemometer heights to 30 feet (9 meters) has been made, many observations are presently made from

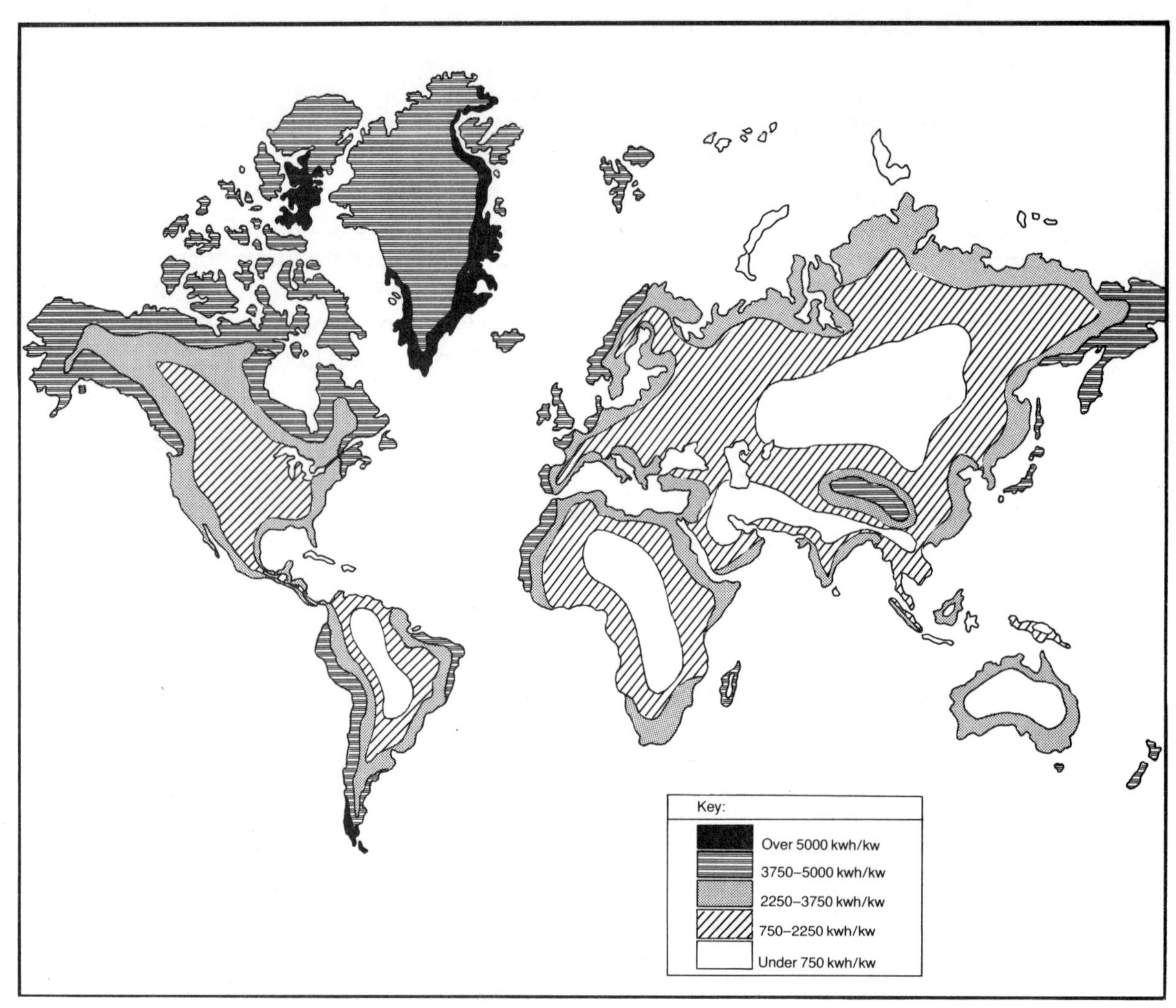

Fig. 2–13: Annual availability of wind energy.

heights of 12 to 60 feet (3.5 to 18 meters), and in the past some readings were taken at heights greater than 180 feet (55 meters). Since wind speeds tend to be higher as the height above the ground increases, the effect of elevation must be accounted for by correction to a standard height before station comparisons can be made.

The earliest anemometers were of the one mile (1.6 kilometer) contact type. The passage of each mile of wind across the anemometer site was registered and the number of miles (kilometers) in a given time period could be converted to miles (kilometers) per hour. By the 1930's aviation interests required wind measurements over a shorter period of time and the anemometers were changed to 1/60 mile (27 meters) contact instruments. The number of 1/60 of a mile (27 meters) contacts in one minute could then be converted to miles (kilometers) per hour.

Between 1905 and 1948 weather stations in the United States recorded the number of miles (kilometers) of wind for each hour of the day. A "5-minute maximum wind speed" or "maximum wind velocity" was also calculated by determining the greatest number of miles of wind recorded

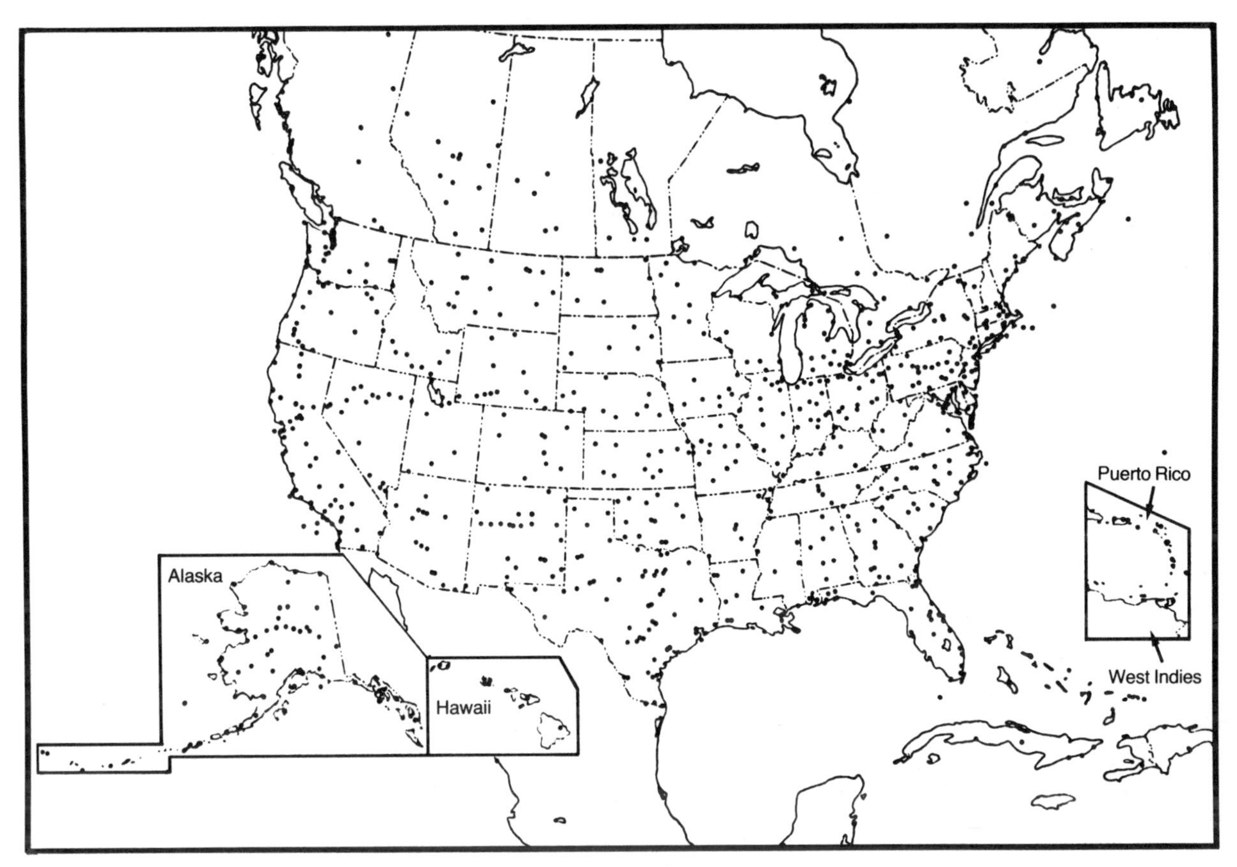

Fig. 2–14: Stations with wind summaries with five or more years of data

in any 5-minute period during each 24-hour period. Beginning in 1912, another measurement, called the "fastest mile" or "extreme wind velocity" was recorded. This was calculated by determining the shortest period of time in which a mile (kilometer) of wind passed by the anemometer, and converting this time to miles (kilometers) per hour. During the 1940's weather stations began to convert the contact anemometers to the magneto type, from which the "fastest mile" and the "5 minute maximum velocity" readings are not readily obtainable. As the number of stations that could furnish these two readings declined, they were gradually phased out of weather bureau reports, particularly after 1957.

The magneto type anemometer yields a continuous record of the instantaneous wind speed on a dial or on a chart recording. As this type of anemometer measures the actual wind speed, gust data can be obtained. The hourly observations of wind speed are made by estimating the average speed for a 1 minute period on the hour by observing the dial or the strip chart recording. A wind gust, when observed, is noted on the form for reporting meteorological observations. A gust is a short duration wind speed that exceeds the lowest observed speed by at least 10 mph (16 kmph) during the observation period. Some errors in anemometer readings are always present, although the magneto type produces smaller errors than the contact type. Until 1933 all readings entered on reporting forms were uncorrected, and correction tables were required to convert these readings to true values. Since 1933 only corrected values are recorded.

The National Climatic Center summarizes wind information from reporting stations into convenient forms ideal for understanding wind characteristics throughout the country. The usual method of summarizing wind information at the Center is to compare wind speed with direction. Fig. 2–15 is an example of a wind summary with direction expressed in the 16 compass points and wind speed at specific speed intervals. These summaries are prepared routinely to satisfy a requester's requirements, and most summaries cover at least a 5-year period as this is the minimum time usually considered necessary for an accurate summary.

SPEED (KNTS) DIR.	1–3	4–6	7–10	11–16	17–21	22–27	28–33	34–40	41–47	48–55	·56	%	MEAN WIND SPEED
N	.1	.3	1.3	1.1	1.0	.5	.2	.1	.8	.0	.1	4.8	15.8
NNE	.2	.1	.5	.7	.9	.8	.6	.1		.0	.0	3.8	19.8
NE	.2	.1	.6	.8	1.2	.8	.7	.1	.0	.1		4.5	19.4
ENE	.1	.1	.3	.8	1.3	1.5	1.2	1.0	.3	.2	.9	7.7	30.0
E	.2	.4	1.1	1.9	1.5	1.0	.9	1.0	1.0	.3	.2	9.5	23.7
ESE	.1	.1	.5	1.0	1.3	1.3	1.1	.8	.8	.7	.1	7.9	27.4
SE	.1	.3	.6	.7	1.1	1.6	1.3	.9	.4	.3	.1	7.6	25.1
SSE	.0	.2	.3	.2	.3	.4	.4	.7	.5	.8	.3	3.9	32.1
S	.3	.3	.6	.9	1.0	.7	.7	.8	.5	.4	.2	6.4	24.9
SSW	.1	.1	.5	.6	.4	.7	.7	.3	.0	.2	.2	3.7	24.9
SW	.1	.2	.6	1.2	.9	1.1	.9	.5	.3	.2	.2	6.3	23.9
WSW	.0	.2	.8	.9	1.1	1.1	.9	.6	.3	.2	.0	6.1	23.0
W	.2	.3	.7	.9	1.3	1.4	.9	.4	.1	.1		6.3	20.7
WNW	.2	.3	.5	1.1	1.0	1.1	.7	.6	.2	.1		6.0	21.3
NW	.3	.4	1.7	1.9	1.4	1.4	.7	.6	.2	.0		8.7	18.1
NNW	.0	.2	.8	1.7	1.5	1.1	.4	.2	.0	.1		6.0	18.6
VARBL													
CALM												.9	
	2.3	3.7	11.4	16.4	17.3	16.6	12.2	8.7	4.8	3.7	2.1	100.0	22.9
								TOTAL NUMBER OF OBSERVATIONS					22.9

Fig. 2–15: Percentage frequency of wind direction and speed from hourly observations (Amchitka Island, Aleutian Islands) for the month of January

How Meteorological Records Are Used in Evaluating Wind Power Sites

The state of Kansas is near the center of the U.S., and in the southwest part of the state is Dodge City. Approximately 175 miles (282 kilometers) to the northeast is the city of Concordia. Without visiting either city, their potential as wind power sites can be compared from weather records.

Fig. 2–16 compares the number of hours per year of winds up to 20 mph (32 kmph) for the stations at Dodge City and Concordia. The figure shows that Concordia has over 1100 hours of wind with a velocity of 6 mph (10 kmph), while Dodge City has less than half as many hours at this speed. For a wind speed of 10 mph (16 kmph) both cities have about 700 hours per year, while for wind speeds of greater than 10 mph (16 kmph) Dodge City has many more productive hours per year.

To the average person viewing this graph it is probably not clear which city has the greater potential as a wind power site. Before continuing, it will be necessary to consider the importance of wind speed in producing power. In mathematical terms the power in the wind varies as the cube of the speed. This relationship is shown graphically in Fig. 2–17, where the values given to the vertical scale are relative, being based on the power at 20 mph (32 kmph) equal to 100.

This graph shows that the power at 20 mph (32 kmph) is 10 times greater than the power at 9 mph (14.5 kmph) and strikingly illustrates the wind power "rule-of-thumb" that it is not feasible to utilize wind speeds of less than 10 mph (16 kmph).

Since the total energy or work available in the wind is a product of the power at various speeds (Fig. 2–17) and the number of hours per year that the wind blows at various speeds (Fig. 2–16), these graphs will produce the data shown in Fig. 2–18. This figure shows the relative wind energy per year at various wind velocities for the two sites. The total available energy in the wind for speeds above 9 mph (14.5 kmph) is exactly twice as much for the location at Dodge City than the site at Concordia. But only 7 percent of the total energy in the wind exists in speeds below 9 mph (14.5 kmph) at Dodge City, while 17 percent exists at Concordia. A wind generator is far more likely to prove satisfactory and economical at Dodge City than at Concordia on the basis of the information contained in this graph.

Fig. 2–16: Comparison of wind patterns in Concordia and Dodge City.

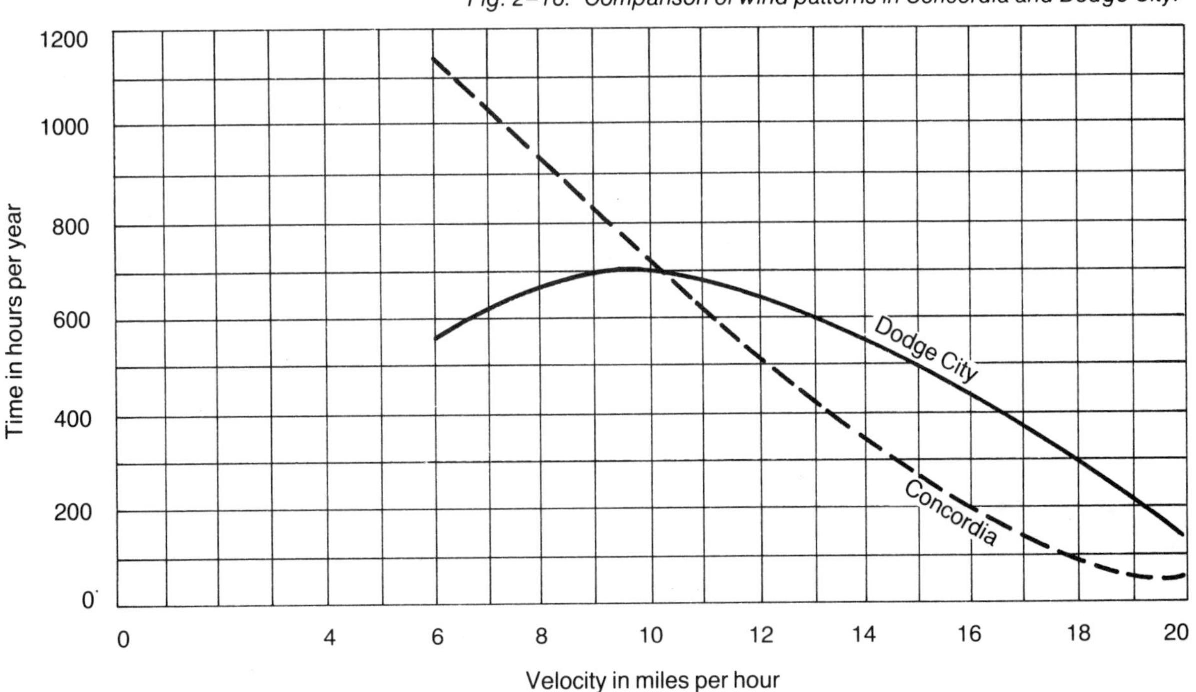

Another important factor which determines the usefulness of winds for generating electrical energy is the distribution of the winds throughout the year. From weather records the distribution of useful winds for each month at the two cities is compared in Fig. 2–19. Although this graph shows that the most favorable months are March and April, it also indicates that the general distribution throughout the year should be satisfactory.

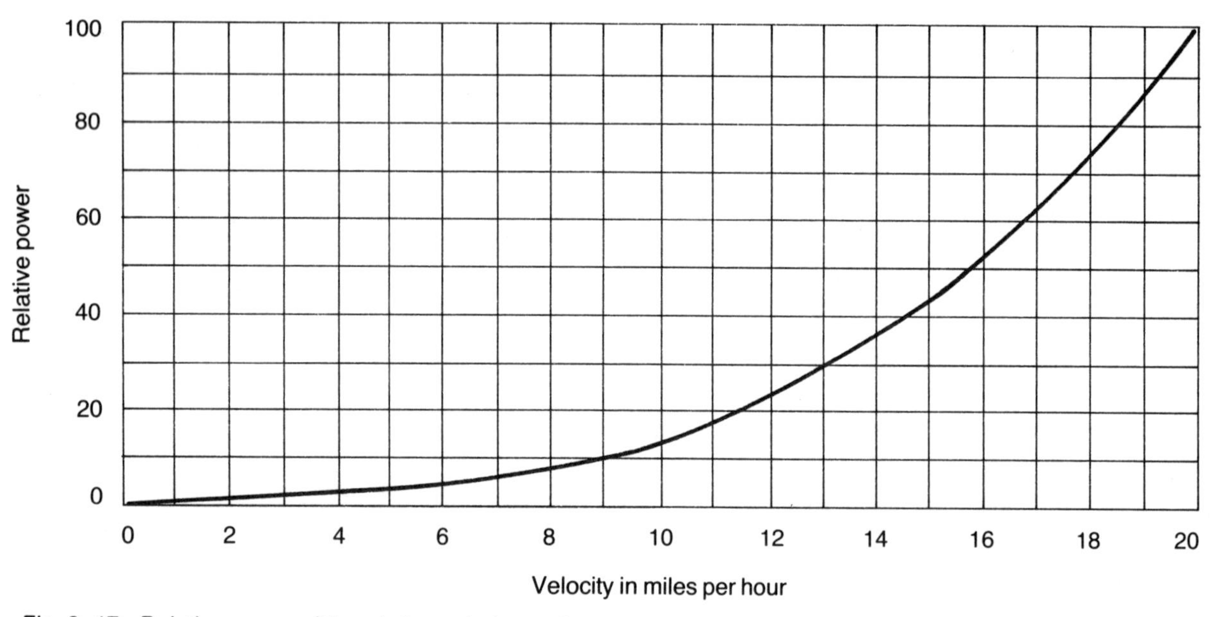

Fig. 2–17: Relative power of the wind vs. wind speed.

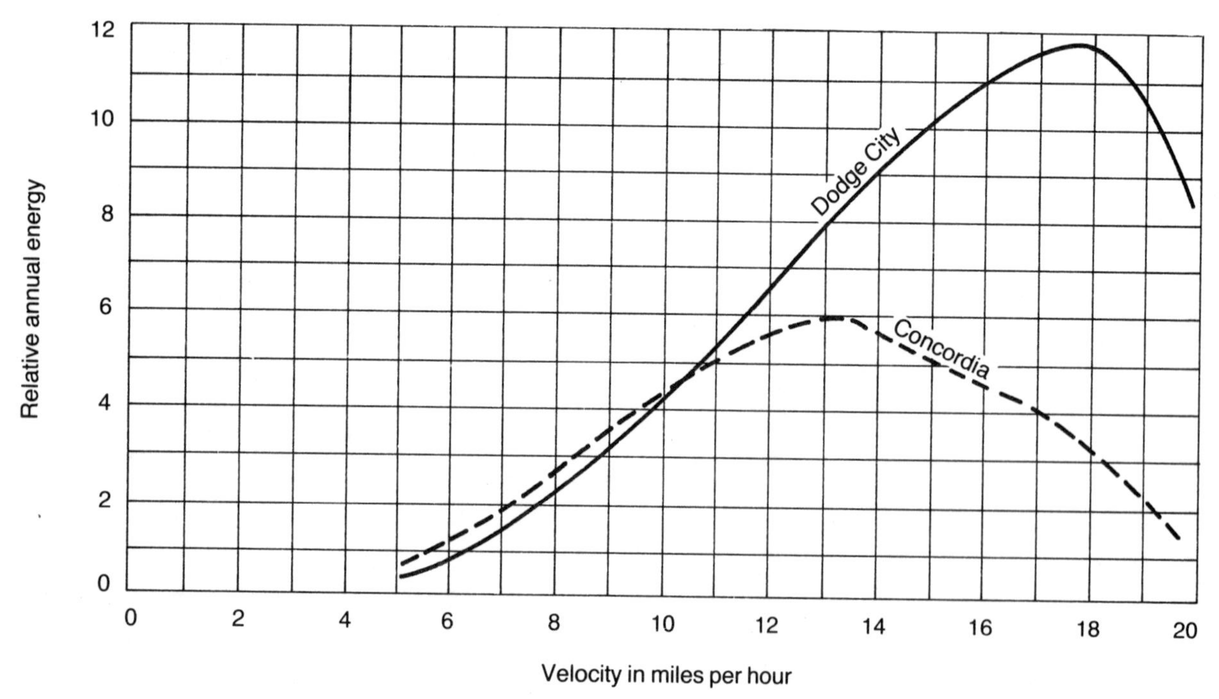

Fig. 2–18: Relating available wind energy per year at varying wind velocities for Dodge City and Concordia.

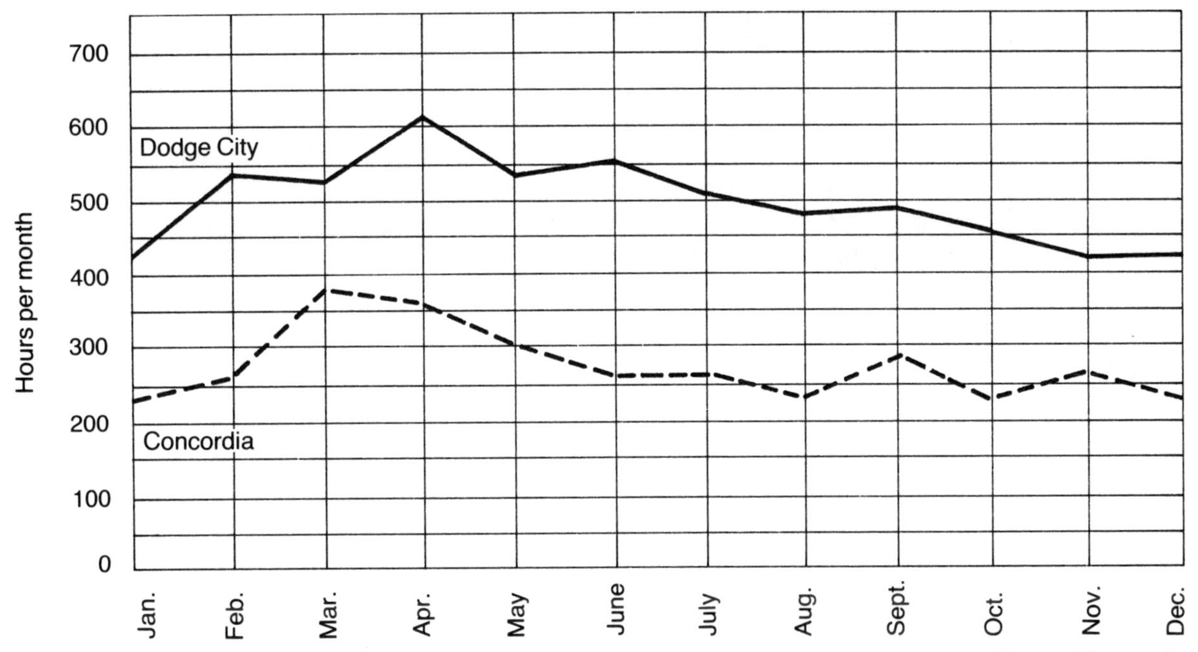

Fig. 2–19: Distribution of useful winds for each month at Dodge City and Concordia.

The number of consecutive calm days is important in evaluating potential wind power sites. This information is readily obtainable from weather records and aids the designer of a wind generator installation in determining the size and capacity of the storage system. For wind generator installations a calm day is usually defined as one in which the wind speed never reaches 10 mph (16 kmph). Thus, little useful power could be generated on such a day. Data compiled from weather records for Dodge City and Concordia over a 5-year period are shown in Fig. 2–20. Concordia has a total of 68 days of calm to only seven at Dodge City. Of much more importance is the fact that Dodge City never has a period of two consecutive days or more of calm weather, while Concordia had 10 periods of two consecutive days of calm, three periods of three consecutive days of

		Consecutive days calm* per year				
Locality	**Total days calm* per year**	**2**	**3**	**4**	**5**	**6**
Concordia . . .	68	10	3	1	1	1
Dodge City . .	7	0	0	0	0	0

*A calm day is one in which the wind never reaches the velocity of ten miles per hour in a twenty-four-hour period.

Source: Kansas State College Bulletin #52 "Electric Energy from Wind."

Fig. 2–20: Calm days per year in Concordia and Dodge City

calm, and one period each of four, five, and six consecutive days of calm. Thus, for Dodge City the storage system would be required to supply the total load for only one day before some recharging energy could be expected. For Concordia the storage system should be of sufficient capacity to supply the total load for six days without significant recharging winds. This information shows that the storage system at Concordia would be much more expensive than that at Dodge City, if total independence from other power sources is desired.

Local Site Selection

Meteorological records are very useful for identifying general locations for wind power sites within large geographical areas, such as a state or county, or within a particular region. But meteorological records will usually provide only limited help in identifying the best wind power sites on a local basis, such as the best sites on a 400 acre (162 hectares) farm or average-sized town.

The chief value of meteorological records in local site selection is to identify prevailing wind directions. Although wind generators will rotate to meet winds from any direction, obstacles on the ground near a wind generator can significantly alter wind speeds and either increase or decrease the flow of wind against the blades of a wind power plant.

How Wind Behaves Close to the Ground

The motion of wind next to the ground produces friction, and this friction causes the wind near the ground to slow down. The amount that the wind is slowed down depends on the roughness of the terrain and the speed of the wind, and will vary at different locations. The slow down at various

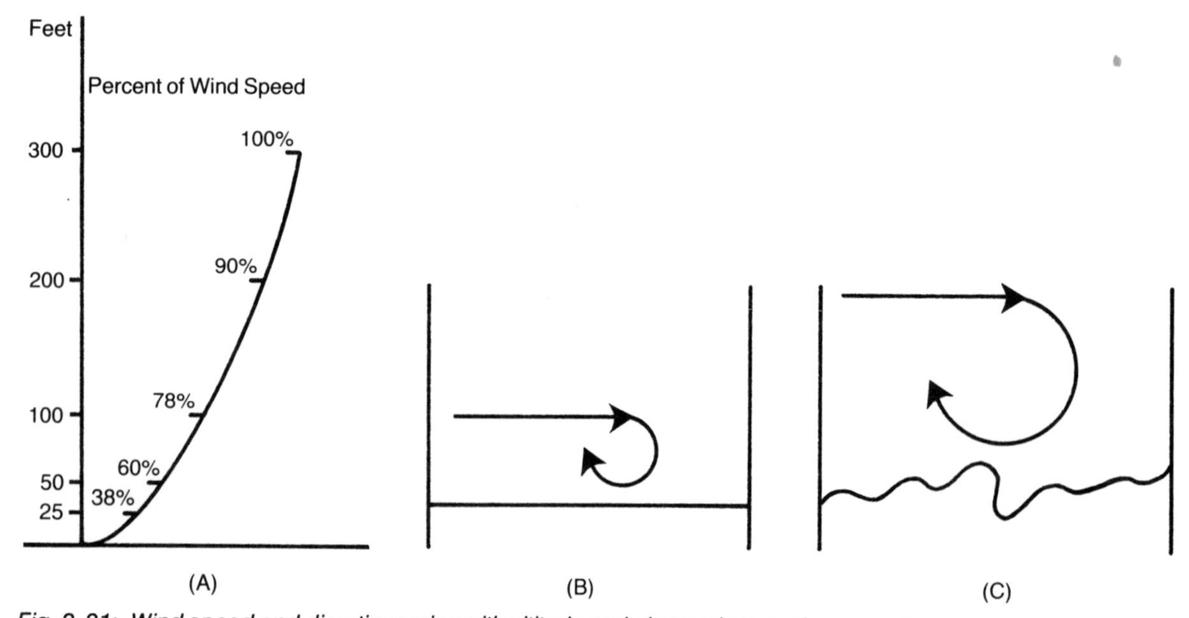

Fig. 2-21: Wind speed and direction varies with altitude and obstructions on the ground.

heights is identified by a wind profile graph (Fig. 2–21A), which shows what percentage of the wind speed at a specified height (in this example, 300 feet [91.5 meters]) is actually available at lower heights above the ground.

In addition, the layer of wind nearest the surface may tumble, causing eddy currents (Fig. 2–21B & C). The size of the eddy currents will vary from less than 1 foot (0.3 meters) with low wind speeds over smooth terrain to more than 50 feet (15.24 meters) at wind speeds of 15 mph (24 kmph) or more over rough terrain. Eddy currents interfere with normal wind flow and reduce the power a wind generator can produce at a particular wind speed.

When wind near the ground flows over obstructions, such a building or steep-sided hill, the normal wind profile is distorted, and transformed into a complicated pattern of reverse flow, eddy currents, and other irregular air movements broadly classified as *turbulence.* Fig. 2–22 illustrates how an obstruction to normal wind flow such as a building causes wind turbulence. This figure shows why the top or lee side of a building is usually a poor site for a wind power plant. If such a site is chosen, the tower should be at least three times as high as the building. This is because eddy currents in front of and behind the building as well as a reverse flow above the top and a low wind speed wake zone on the lee side each contribute to reduced wind speed near the building. In addition to reducing the potential power output of a wind generator, the blades of a windmill situated in these locations will be constantly buffeted by turbulence which will usually result in premature mechanical failure of the wind generator.

The importance of wind power site selection was given further emphasis when, during the early 1940's, it was discovered that certain types of terrain have higher wind speeds than nearby areas. Wind speed measurements over low, smooth hills gave surprisingly higher readings than the surrounding plains (Fig. 2–23). What happens is that locations with trees deformed by the wind will usually be excellent wind power sites.

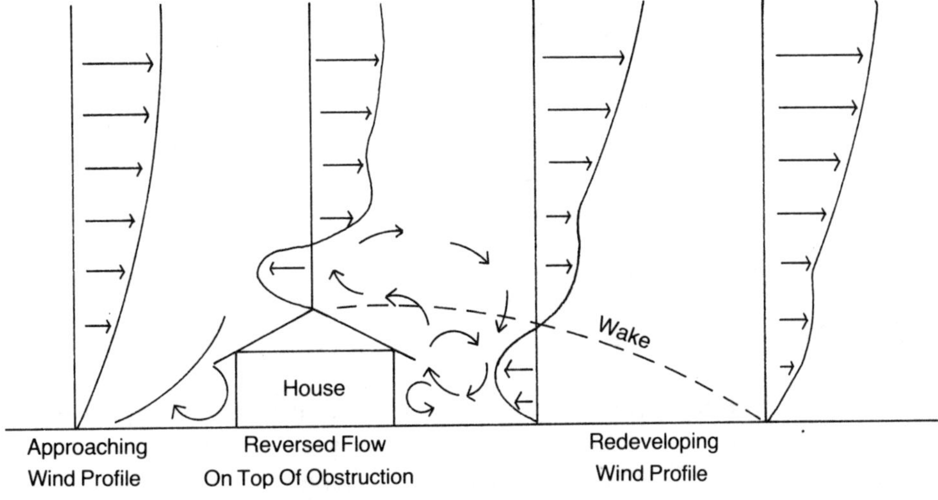

Fig. 2–22: Wind turbulence caused by ground-level obstruction.

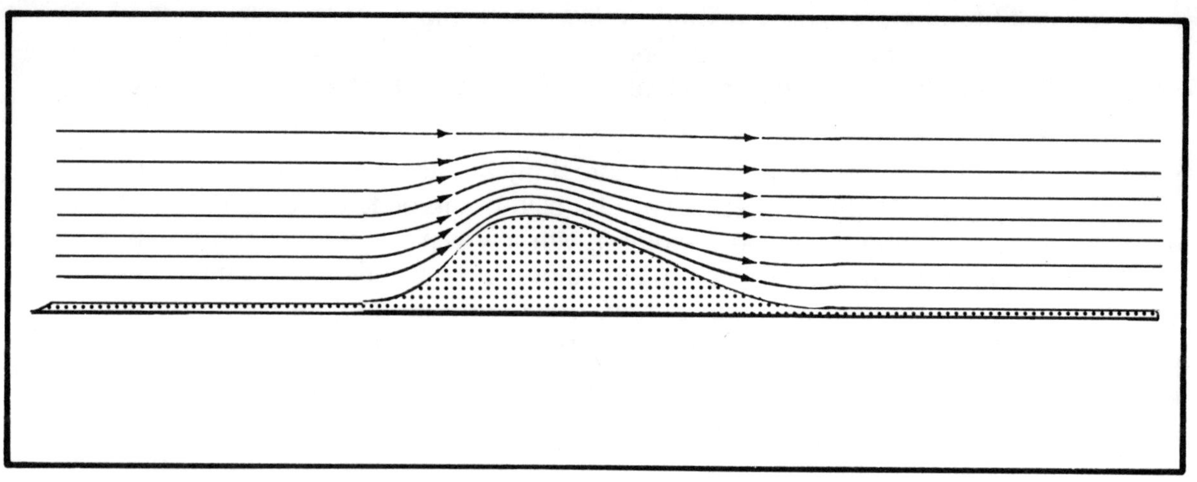

Fig. 2–23: Acceleration of wind over hill.

Increase of Wind Speed with Height

Representative wind profile graphs for different types of terrain (Fig. 2–24) show the effects that surface friction and *obstacle-produced turbulence* have on reducing wind speeds at various heights above the surface. Wind profiles vary with the type of terrain and with the wind speed. Wind speeds at low heights are greatest over water and smooth unbroken plains, less over rural areas with buildings and trees, and still less over metropolitan areas where the close spacing of tall buildings distorts normal wind profiles.

In this representative example, a wind generator on a 50-foot (15-meter) tower situated on a platform or barge on a large lake would capture 78 percent of the wind speed available at 200 feet (61 meters). The same generator and tower sited in a rural area with farm buildings and trees would receive 55 percent of the wind speed at 250 feet (76 meters), while the wind speed at 50 feet (15 meters) in a metropolitan area would be reduced to only 35 percent of that available at 300 feet (91 meters). As the size of ground obstacles increases, the height of the gradient wind also rises.

For most wind power installations the type of terrain is the most important factor in deciding whether the increased cost of a very high tower is justified. In addition to capturing a greater percentage of the gradient wind, raising the tower height of the generator above the most turbulent wind layers greatly prolongs the life of the machine.

Wind Measurements

By now you probably have a good idea of where the best wind power sites in your area are likely to be found—you may even have a few specific locations in mind. The next step is to take a series of wind measurements at each site. These measurements can then be compared to identify the

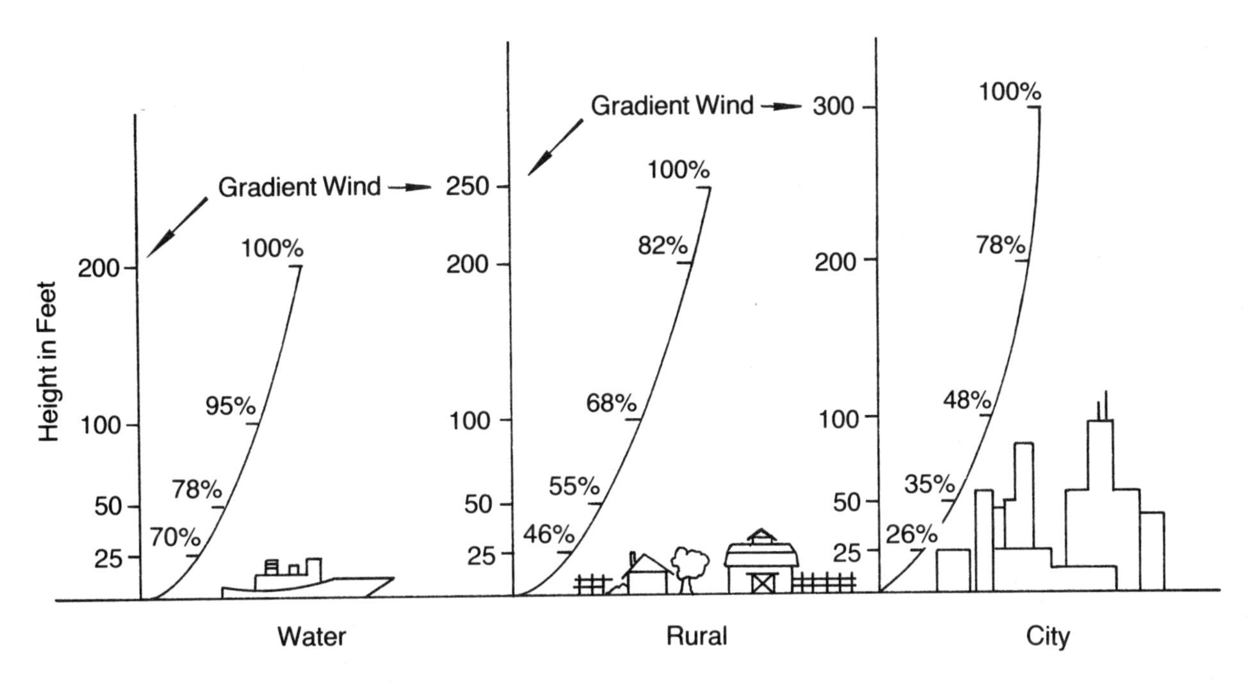

Fig. 2–24: Wind profile graphs showing effects of obstacle-produced turbulence.

site with the greatest average wind speed. Even with only one site you can compare the readings from this site with wind records available from the nearest weather station. This may be a station that supplies records to the National Climatic Center, or the nearest airport, military installation, government agency, or university. Many public schools and private colleges maintain weather stations as a teaching aid and can supply wind records for many years (Fig. 2–25).

A number of wind speed measuring instruments are available at prices from less than 10 dollars to several hundred dollars. The Dwyer Instrument Company of Michigan City, Indiana, manufactures an inexpensive hand-held wind meter and a remote-reading pneumatic wind speed indicator complete with 50 feet (15 meters) of flexible tubing (Fig. 2–26). These instruments as well as electric anemometers that measure winds to 100 mph (161 kmph) (Fig. 2–27) are available from Edmund Scientific and the American Science Center (see Source/Resource section) All but the hand-held instrument can be mounted at practically any height.

Better still is a recording anemometer that registers the total amount of wind that passes its position. A reading of the anemometer counter before and after any period—a day, week, or month—can be readily converted to the average wind speed during the period. But recording anemometers are considerably more expensive than measuring anemometers and the two Dwyer wind speed indicators. By taking wind speed measurements as often as possible at the same height at each prospective site for at least three months one should have enough information to select the best site, or to determine whether a particular site has more or less wind than the nearest weather station. Only by following this procedure can we avoid future disappointment once a wind generator is placed in service.

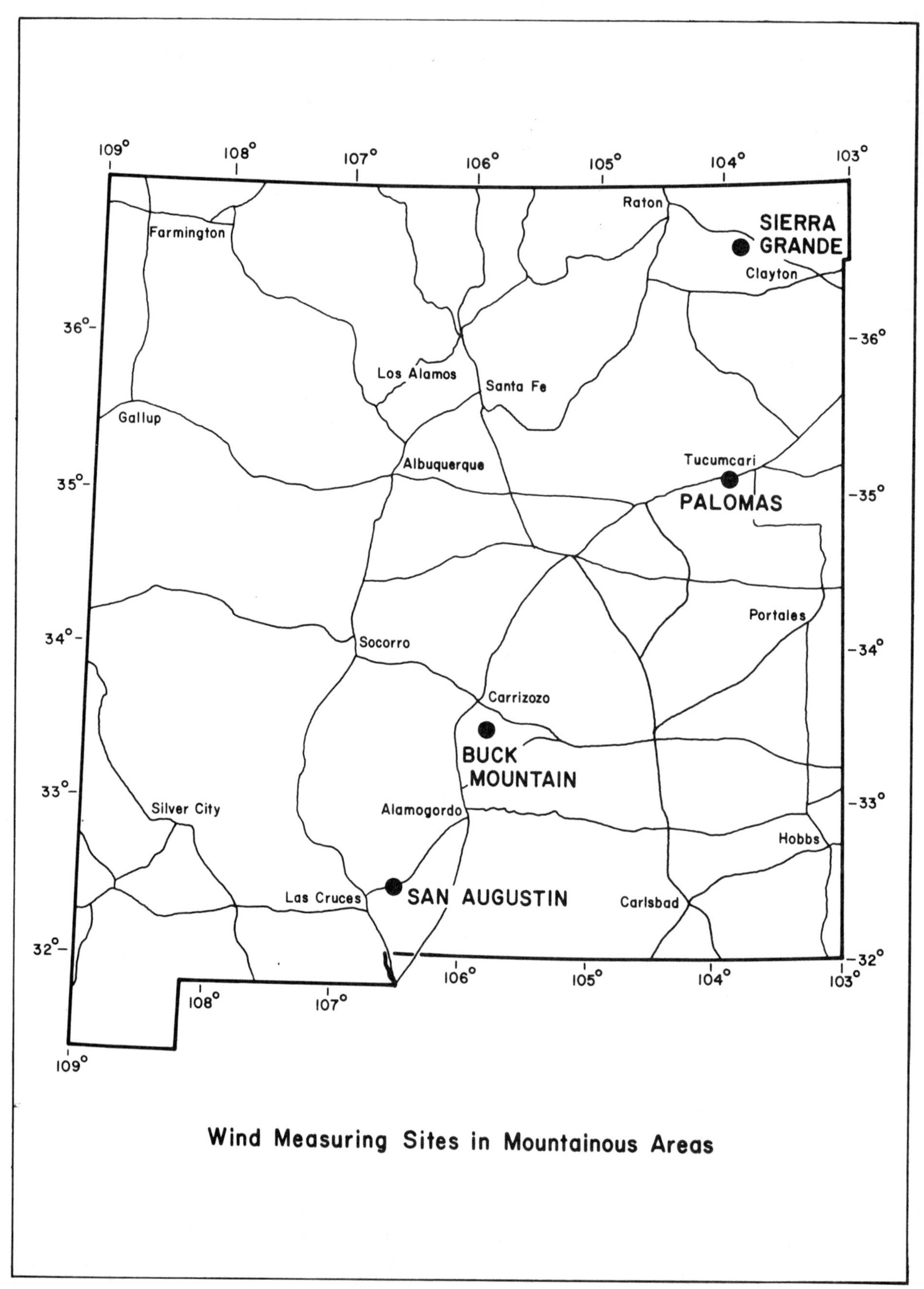

Fig. 2–25: An example of measuring sites in New Mexico.
(Similar maps can probably be gotten from wind/meteorological stations in your area).

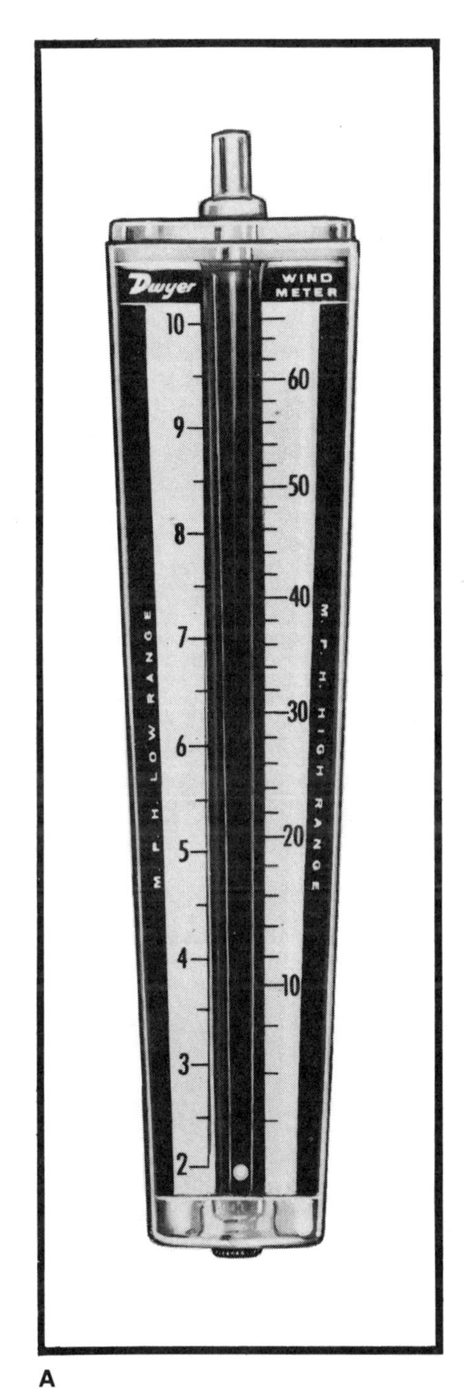

A

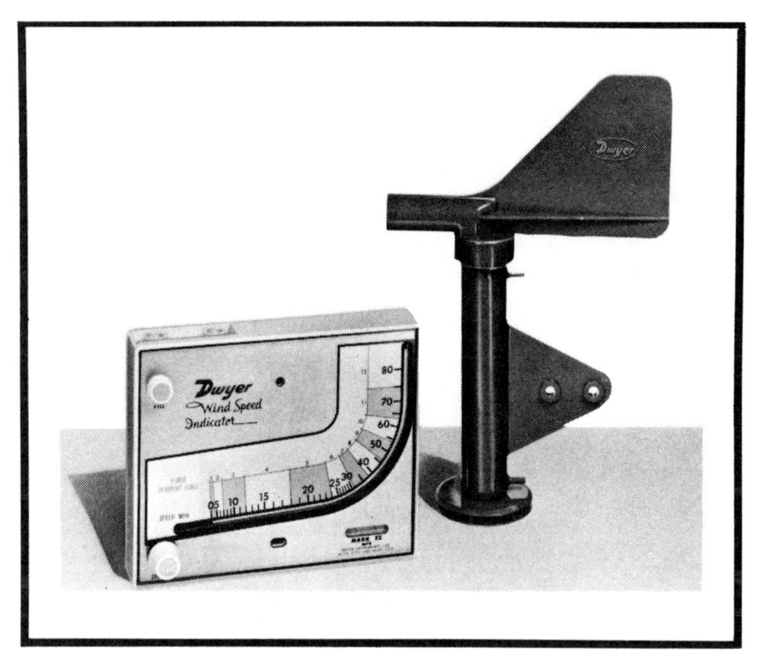

B

C

A:
Handheld wind meter.
(Dwyer Instruments Inc.)

B:
Wind speed indicator Mark II.
(Dwyer Instruments Inc.)

C:
Anemometer Model RK-7.
(R.A. Simerl Instrument Division)

A: Handheld (cup) anemometer.
(R.A. Simerl Instrument Division)

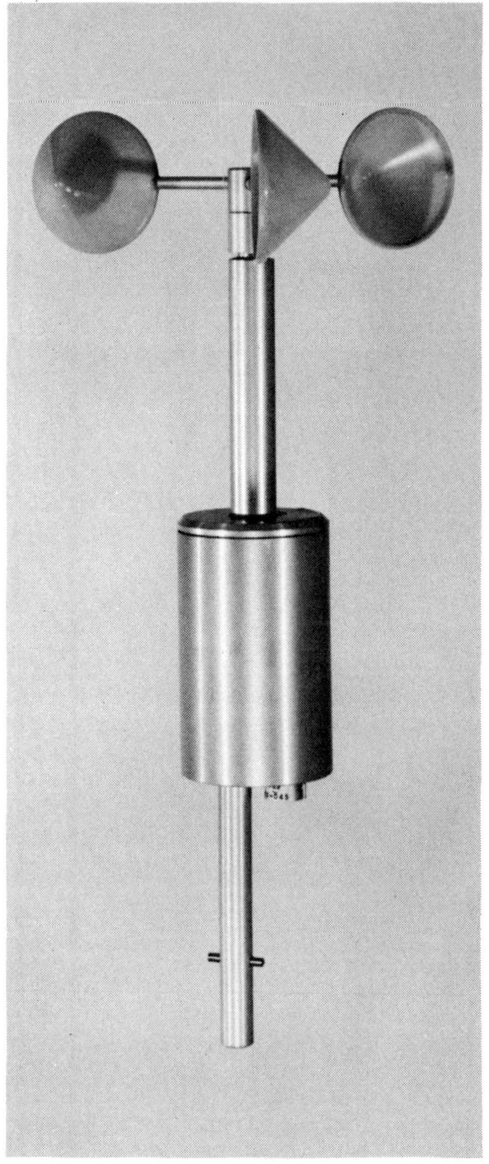

B: Post-mounted (cup) anemometer.
(Climet Instruments Company)

SUMMARY OF WIND GENERATOR SITE SELECTION

The following guidelines summarize the most important steps in selecting a site for a wind generator:

1. Whenever possible, study meteorological records for the general area where you plan to install a wind generator. If an official weather station is not nearby, such as those in the United States that supply wind records to the National Climatic Center, check at the nearest municipal or private air field, military installation, government agency, or university. Other sources of wind records throughout the world include many public schools and private

colleges. When studying weather records, pay particular attention to the prevailing wind direction, average wind speeds throughout the year, daily peak wind speeds, and the number of days of calm, or winds of less than 10 mph (16.1 kmph).

2. Closely examine the terrain where you plan to install a wind generator. Select trial locations with the prevailing wind direction in mind, and at least 300 feet (91.5 meters) from obstacles such as buildings and trees. Observe how vegetation such as trees are deformed by local wind patterns. Look for low hills with gentle slopes where a wind speed-up factor may be present.

3. Conduct wind measurements at each trial location and compare these readings with wind records from the nearest weather station. The accuracy of your measurements will increase as the length of time that measurements are made increases. Three months of wind measurements is usually considered the minimum; but, if possible, obtain wind speed measurements at each trial location for every month of the year.

   With remote-reading instruments try to take wind measurements at least 20 feet (6 meters) above nearby obstacles. Use a hand-held wind meter only in open terrain away from buildings and trees. Wind measurements with even the simplest instruments are better than none at all and will provide useful information that can be compared with nearby weather stations to determine whether the trial locations selected have above or below normal wind speeds.

Selecting the best available site for a wind generator is the single most important factor in planning a wind power installation and will often determine whether the venture will be an economic success or failure.

There may soon be a day when there are as many of these as there are TV antennas! (Enertech Corporation)

Chapter Three

THE GENERATION OF ELECTRICITY BY WIND POWER

Electricity, that magic genie ready to work at the flip of a switch, is one form of energy that can be readily produced by wind power. Mechanical energy is another form of energy produced by wind power, but it is often less useful than electrical energy. Generally, mechanical energy must be used when and where it is produced. Electrical energy can be connected to a distant load by wires, and one type of electricity can be stored for future use.

It may help, at this moment, to think of the water pumping windmill, and the importance of site selection. The water pumping windmill converts wind power to mechanical energy to pump water from a well beneath the windmill tower. Since the windmill and the well are located together, the windmill may have to be placed at a poor wind power site if the location of the well is of utmost importance. But, if the windmill were placed at the best site for wind power the well may be in a poor location for water distribution.

With a windmill generating electricity, the wind generator can be placed at the best wind power site, and the well and an electric water pump can be located wherever they are needed. The wind generator and the pump (i.e. the load) are then connected by wires and the correct control devices. With a wind generator, an electric pump, and storage batteries, it is possible to store enough electricity to operate the pump when the wind is not blowing.

Comparing windmills that produce mechanical energy with those that generate electricity may appear to be a simple matter, but when a steady supply of water is necessary the correct choice is of utmost importance. Recent studies in the Soviet Union have resulted in a recommendation for the use of wind generators and electric water pumps in arid regions of Russia rather than the simpler mechanical water pumpers. The Russian studies concluded that one medium-size wind generator could operate five or six electric water pumps at widely spaced wells. Even if a few wells went dry or became contaminated with salt water (a problem in the region studied) the others would still supply a sufficient amount of fresh water.

Since the future use of wind power is closely linked to wind generators, one needs to have a basic understanding of electricity. For example, electricity produced from wind power may or may not be the same type of electricity at work in your home. This is because not all wind generators produce the same type of electricity.

A detailed knowledge of electricity is not needed to understand wind generators, nor is such knowledge necessary to understand home wiring and electrical appliances. Only a few fundamentals are absolutely necessary, but should you decide to include a wind generator in your home wiring system a more thorough knowledge of electricity will be necessary. To this, add an understanding of electrical safety practices. Also, you will need to check local regulations governing electrical work in your area. In many locations building codes require electrical work to be done or inspected by a licensed electrician. In addition, you should request a review of your home owner's fire insurance policy. Some policies require that all electrical work must be done by a licensed electrician or contractor.

The electricity in your home is most likely 120-volt alternating current, or AC. Voltage is a measure of electrical pressure, or potential. Wiring of 120-volts has 10 times the potential of 12-volts, and one-half the potential of 240-volt wiring. Electrical current is measured in units called *amperes,* or *amps.* Alternating current (AC) is simply the result of how the electricity was generated at the power plant. Huge rotating machines produce electricity that changes direction, or alternates, at a rate linked to the speed of rotation of the machine. In most parts of the United States alternating current of 60 cycles per second is produced. In other parts of the world 50 cycles per second is the standard frequency.

Alternating current can be stepped-up or stepped-down to higher or lower voltages by the use of an electrical device called the transformer. Certain appliances, such as electric clocks, fluorescent lights, and entertainment equipment such as some types of record and tape players, and television and radio receivers can only be operated on alternating current of a specific frequency (50 or 60 cycles) as they are dependent on the frequency of the alternating current for proper operation.

The *ampere*, again, is a measure of the current flow in the wires. Another unit, the *watt,* is used to measure electrical power consumption in the home. One watt is equal to one volt of electrical pressure at one ampere of current flow. The most common measure of watts is the *kilowatt,* which is equal to 1000 watts. The electric meter that records the amount of electricity used in the home measures *kilowatt-hours,* which is one kilowatt of electricity consumed for one hour. Household electric bills are based on the number of kilowatt-hours consumed during each billing period. For example, a 100-watt light bulb burning for 10 hours will use one kilowatt-hour of electricity—100 watts multiplied by 10 hours equals 1000 watt-hours or one kilowatt hour. You may want to read your electric meter on a daily or weekly schedule to determine the energy usage in your home. This is very important if you decide to install a wind generator. Ideally, an energy survey conducted over a full year will give a record of energy use and will help determine the size of wind generator that is able to supply all or part of your electrical power needs.

An electric meter (Fig. 3–1) records the use of kilowatt-hours of electricity. To interpret it:

1. Read the dials from left to right. (Note that the numbers run clockwise on some dials and counter-clockwise on others.)
2. The figures above each dial show how many kilowatt-hours are recorded each time the pointer makes a complete revolution.
3. If the pointer is between numbers, read the smaller one. (The 0 stands for 10.) If the pointer is pointed directly at a number, look at the dial to the right. If that pointer has not yet passed 0, record the smaller number; if it has passed 0, record the number the pointer is on. For example, in Fig. 3–1, the pointer on the first dial is between 0 and 9—read 9. The pointer on the second dial is between 5 and 4—read 4. The pointer on the third dial is almost directly on 5, but the dial on the right has not yet reached 0 so the reading on the third dial is 4. The fourth dial is read 9. You would read the third dial as 5 after the pointer on the fourth dial reaches 0. Thus, the total reading is 9,449.
4. This reading is based on a cumulative total—that is, since the meter was last set to 0, 9,449 kilowatt-hours of electricity have been used. To find your daily or weekly use, take readings each day or week and subtract the earlier, smaller reading from the

Fig. 3–1: Interpreting an electrical meter.

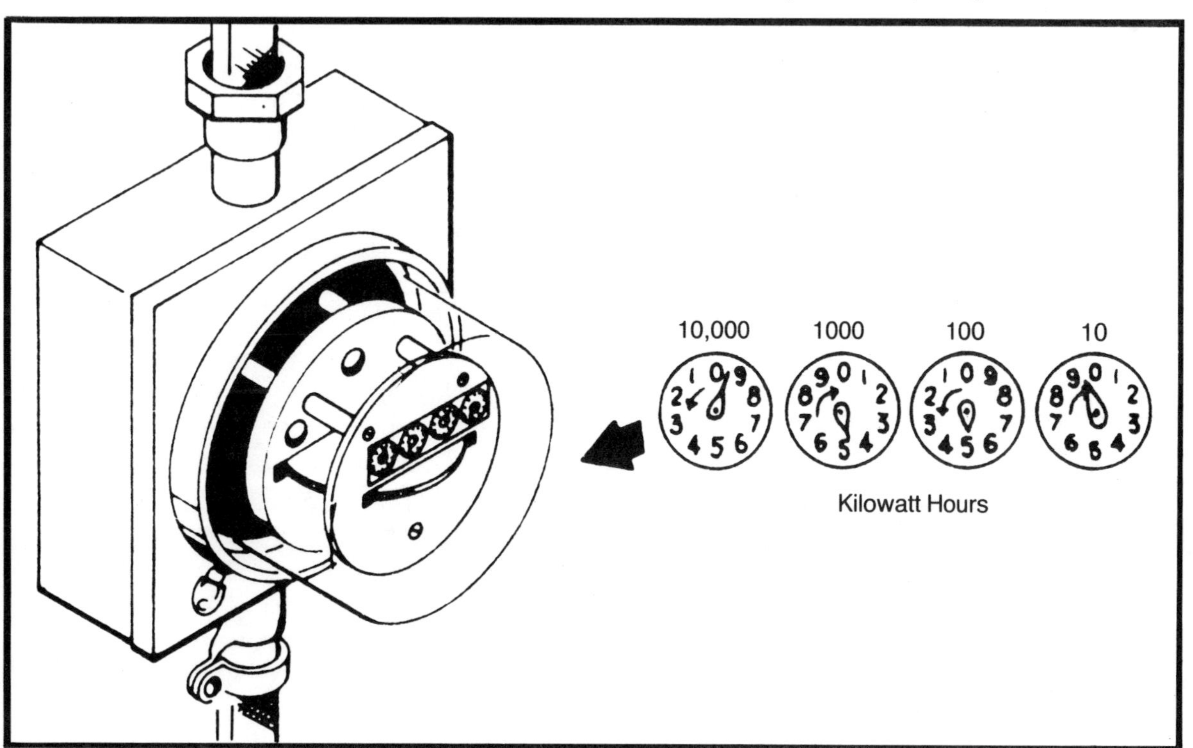

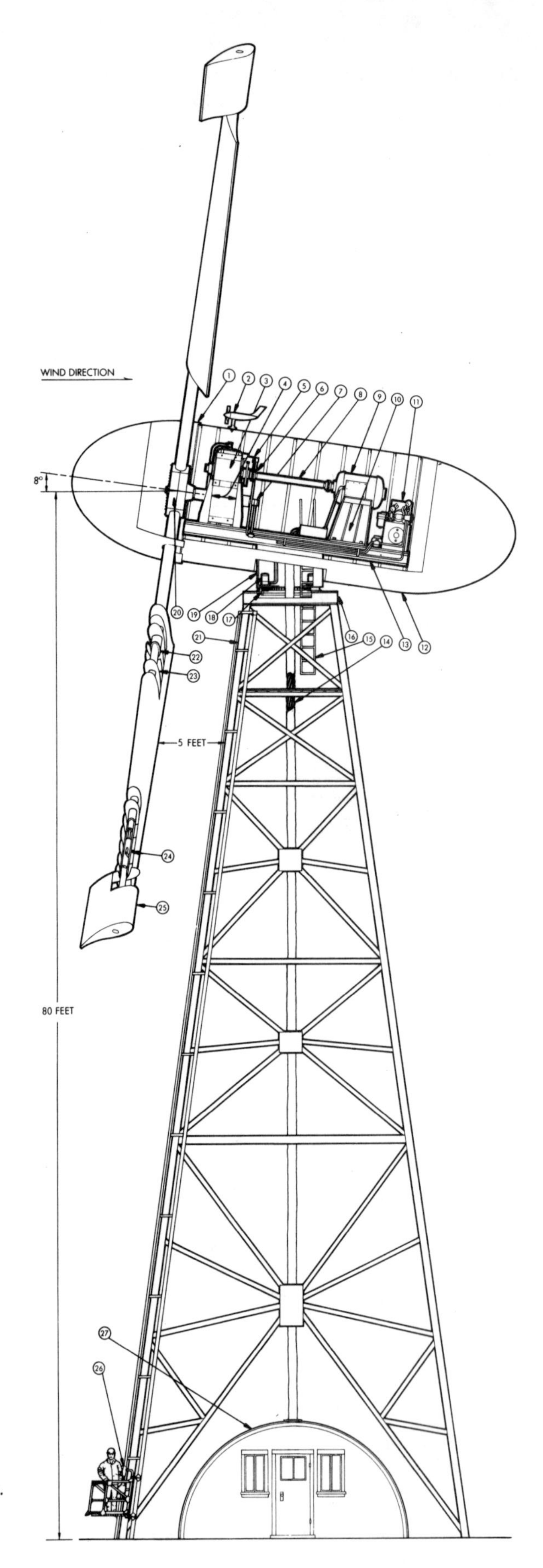

DETAILED SCHEMATIC OF THE MP1-200 PRODUCTION PROTOTYPE ERECTED ON CUTTYHUNK ISLAND

1 SPINNER, GRP
2 WIND SPEED AND DIRECTION SENSOR
3 GEAR TRANSMISSION
4 MAIN BEARINGS
5 DISC BRAKE
6 FLEXIBLE COUPLINGS (2)
7 ROTATING HYDRAULIC UNION
8 HIGH SPEED SHAFT
9 SYNCHRONOUS GENERATOR, 250 KVA
10 GENERATOR MOUNT
11 HYDRAULIC CONTROL ASSEMBLY
12 MACHINE CABIN
13 MACHINE BEDPLATE
14 POWER AND CONTROL CABLES
15 ACCESS LADDER
16 TOWER CAP
17 TURNTABLE BEARING/BULL GEAR
18 YAW CONTROL SERVO MOTORS (2)
19 TURNTABLE MOUNT
20 ROTOR HUB, FIXED PITCH
21 PINNED TRUSS TOWER, COR-TEN STEEL
22 BLADE SPAR, SEAMLESS STEEL TUBING
23 BLADE RIBS
24 TIP ACTUATING MECHANISM
25 BLADE TIP/DRAG FLAP
26 ELEVATOR
27 CONTROL HOUSE, MICRO-PROCESSOR CONTROL SYSTEM

Detailed Schematic of the MP1–15 Prototype. (WTG Energy Systems Incorporated)

later one. Thus, if the reading on the following day was 9,470 kilowatt-hours, your electrical usage for that 24-hour period would be 21 kilowatt-hours.

From this survey you will undoubtedly see consumption of kilowatt-hours varying in your home with the seasons of the year and with the number of persons living in the home. In addition, you will get a good idea of the amount of electricity used by various appliances in your home. For example, the cost of operating an air conditioner, washing machine, and an electric clothes dryer will be readily evident.

A DIFFERENT TYPE OF CURRENT

Not all wind generators produce 120-volt alternating current. Indeed, many smaller units generate direct current, or DC. Direct current cannot be stepped-up or stepped-down to higher or lower voltages by transformers, and many appliances designed for 120-volt alternating current will not operate on direct current. But, direct current can be stored in batteries. And, modern electronic devices called inverters can be used to step-up battery voltage and change it to alternating current. The smallest and least expensive wind generators produce direct current at 12-volts, but larger units may produce current at 24-, 36-, or even 120-volts DC.

Wind generators that produce 120-volt DC power can be used to operate many appliances that will function equally well on either AC or DC power. These include incandescent light bulbs; appliances that use heated wires or rod electrodes such as water heaters, toasters, and electric broilers and ranges; and devices powered by universal brush-type electric motors, such as portable electric drills, mixers, vacuum cleaners, and some types of pumps, blenders, and electric fans. A small DC to AC inverter can be used to supply alternating current to frequency-dependent appliances such as electric clocks and AC-operated entertainment equipment (Fig. 3–2).

Fig. 3–2: DC system with battery storage.

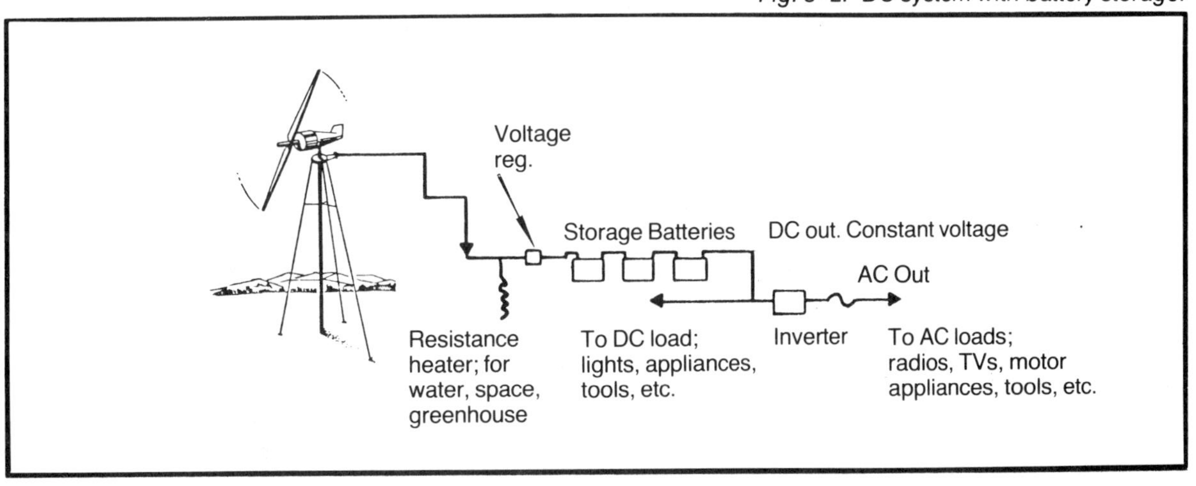

In many wind power installations where the wind generator produces 120-volt DC the wiring in the home is a dual system with separate outlets supplying DC and AC. The outlets are clearly labelled, and the home owner simply plugs each appliance into the outlet supplying the right type of current. The use of DC power outside the home is also practical today. Direct current motors can replace AC motors on well pumps and other machines, and there are available today, from several manufacturers, small electric tractors for farm, garden, and homestead use. These machines are powered by a direct current motor and conventional wet-cell batteries totalling 36-volts. The power of these tractors compares to that of a 16 horsepower gasoline-engine tractor. One company that manufactures these tractors, General Electric, had the foresight to offer a large number of accessories for their machine, the Electrac. In addition to typical tractor-mounted attachments such as plows, grass cutters, rotary tillers, cultivators, and others, they offer a selection of portable accessories that operate from the tractor batteries. These include sprayers and portable tillers, trimmers, drills, chain saws, and others. They also offer a tractor-mounted inverter to produce alternating current. Although tractors such as the Electrac are not full-size agricultural machines, they have proven to be able performers on the type of small farm found the world over where a moderate amount of different crops are grown for market and for subsistence living.

Tractors such as the Electrac with its accessories are an example of a total energy system that would prove ideal for the output of a small or medium-size wind generator producing DC power. Wind generators are an ideal source of power for farms and rural areas where electric tractors can be best utilized. During the night when the home power demand is minimal, the wind generator can be used for charging the tractor batteries.

Distance limits how far direct current can be sent through wires without excessive losses. This is especially pronounced with low-voltage DC, where large voltage losses result when long spans of wire are used. Since DC cannot be stepped-up to higher voltages by using simple transformers, it must be sent over wires and used at the same voltage that is produced by the wind generator. If a particular wattage is required by the work load, the current required at low generator voltages becomes very

Table 3–1. Minimum Size Wire Required

			AWG (American Wire Gauge)	
Watts	=	Amperes × Volts	Copper	Aluminum*
1200 watts	=	100 amps at 12 volts	4	2
1200 watts	=	50 amps at 24 volts	8	6
1200 watts	=	33.3 amps at 36 volts	10	8
1200 watts	=	10 amps at 120 volts	14	12

*Aluminum wire has approximately 78 percent of the current carrying capacity of the same wire size in copper.

NOTE: The larger the AWG rating, the smaller the diameter of the wire.

high. This requires use of larger size cables to avoid excessive losses. (See Table 3–1 below.)

Thus, 1200 watts can be supplied by smaller and less expensive wires if it is 10 amperes at 120-volts of electrical pressure. The same wattage supplied from a 12-volt wind generator would require 100 amperes and very expensive heavy cable.

The property of wiring that becomes important at low voltages is the *resistance* of the wires. Resistance, or opposition to current flow, is a property that all materials have. Materials such as rubber and plastic have a high resistance to current flow, and are called *insulators.* Materials such as copper and aluminum have a very low resistance to current flow, and are called *conductors.* Resistance is measured in units called *ohms,* and the resistance of wire is usually specified in ohms per 100 or 1000 feet (305 meters) of conductor.

The resistance in wires gives a voltage drop, or loss, over the length of wire that varies with the amount of current passing through the wire. This is given by a simple but invaluable formula called Ohm's Law:

Voltage (drop or loss) $=$ current $\times$ resistance

Fig. 3–3: DC wind generator connected to load.

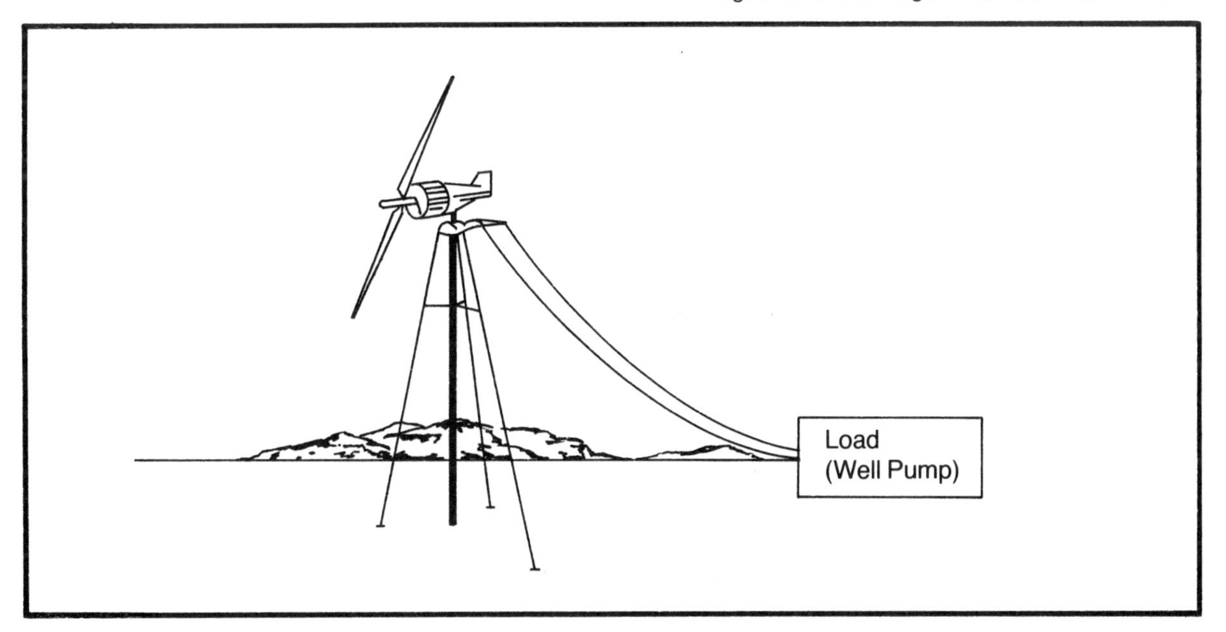

Fig. 3–3 shows a DC wind generator that is supplying 1200 watts of power to a load through 100 feet (30.5 meters) of No. 4 AWG copper wire. The wire resistance is 0.025 ohms. If the wind generator is supplying 1200 watts at 120-volts and 10 amperes the voltage drop or loss across the wire is:

Voltage drop $=$ 10 amperes $\times$ 0.025 ohms $=$ 0.25 volts

This value is very small and will not affect the load. A voltage drop or loss of this small value is negligible and may be ignored in calculations. Indeed, a wire as large as No. 4 AWG would not be necessary, and wire several sizes smaller would serve as well.

But, if the 1200 watts were supplied at 12-volts and 100 amperes the voltage drop would be:

$$\text{Voltage drop} = 100 \text{ amperes} \times 0.025 \text{ ohms} = 2.5 \text{ volts}$$

This value is much larger in proportion to the generator voltage of 12-volts. In this example the 2.5-volt drop or loss across the wire means that only 9.5-volts would be supplied to the load. This would affect the normal operation of the load, which in this example is a well pump with a 12-volt DC motor.

In addition, large power losses take place in the wires. In this example, 250 watts of power are lost in the wire, and only 950 watts are delivered to the load. Since the pump motor may not receive enough power to operate properly it may burn out.

Table 3–2, below, lists the approximate resistance of 100-foot (30.5-meter) lengths of the most common sizes of copper and aluminum wire.

Table 3–2. Resistance in Ohms per 1200 Feet (365.76 meters)

AWG. No.	Copper	Aluminum
14	.265	—
12	.162	.26
0	.100	.16
8	.064	.10
6	.040	.07
4	.025	.04
2	.016	.025

HOW, WINDMILLS CAPTURE WIND ENERGY

Windmills produce mechancial energy which in turn is changed to electrical energy by a generator. How much energy can be produced by a windmill of a certain size at a given wind speed can be found by formulas which are shown in this section. You do not need to understand the derivation of these formulas to use them, as only simple mathematics is required. But, the following paragraphs will greatly increase your understanding of wind power.

Energy in the Wind

The theoretical energy available in the wind can be calculated by the laws of mechanics. Any mass, including air, when in motion has a *kinetic energy* expressed by the relation:

$$KE = \tfrac{1}{2}mv^2$$

where *m* is the *mass* of the air and *v* is the *velocity* of the wind.

The mass of air that actually supplies energy to a wind machine is proportional to the cross-section area (A) covered by the rotating blades of the windmill. For the revolving blades of a horizontal-axis wind generator this is a circular section given by the relation:

$$A = \left(\frac{\pi D^2}{4}\right)$$

where $\pi = 3.1416$ and D is the diameter swept by the windmill blades. Thus, the kinetic energy available in a circular section of wind can be expressed as:

$$KE = \frac{1}{2}m\left(\frac{\pi D^2}{4}\right)v^2$$

Input Wind Power to the Windmill

Translating the kinetic energy present in a circular section of wind to useful work, or power, which can be done by the invisible tube of wind in one second times the distance, v, through which it moves:

$$\text{Power} = (KE)(v) = \left[\frac{1}{2}m\left(\frac{\pi D^2}{4}\right)v^2\right](v)$$

$$\text{Power} = \frac{1}{2}m\left(\frac{\pi D^2}{4}\right)v^3$$

Power Extracted from the Wind

The expression above for the power in a wind column cannot be applied for the continuous extraction of energy from the wind because it assumes that the wind speed is changed from a velocity, v, to rest, or zero. If the air behind the windmill blades is brought to rest or stopped there would be no space available for the column of air which would arrive next and in succeeding moments. It is necessary to allow the wind to pass through the blade area and extract energy by reducing the speed of the wind. In 1927, A. Betz, of Gottingen, Germany, demonstrated that the maximum fraction of power that could actually be extracted was $^{16}/_{27}$, or 59.3 percent, of the total power in the wind column. Thus,

$$\text{Power} = \frac{.593}{2}m\left(\frac{\pi D^2}{4}\right)v^3$$

Taking air density as 0.08 lb per cubic foot, power as watts, cross-sectional area (A) in square feet, and wind velocity in miles per hour (mph), the above equation can be simplified to:

$$\text{Power (watts)} = 0.00314\,AV^3$$

Air density will vary only slightly, being lower with increasing humidity or content of water vapor, and lower also with increasing altitude or height above sea level. Since air density will also vary over any time period with changes in barometric pressure and temperature, an average air density is usually applied to wind power calculations.

The above equation can be further simplified to express the swept area (A) of the windmill in terms of the blade diameter in feet:

Formula 1: $$\text{Power (watts)} = 0.00246D^2V^3$$

1. The power extracted from the wind increases with the square of the diameter of the area swept by the blades.
2. The power extracted from the wind increases with the cube of the wind speed.

The importance of selecting the windiest possible site is readily seen from these expressions. Although doubling the diameter of the windmill blades increases the power available from the wind four times, a doubling of the wind speed increases the power eight times.

Horsepower

Formula 1 above expresses power in terms of watts. The power that a windmill develops also can be expressed in horsepower. Formula 2 below expresses power in horsepower where D is the blade diameter in feet and V is the wind velocity in miles (kilometers) per hour:

Formula 2: $$\text{Power (Horsepower)} = 0.0000033D^2V^3$$

There are more zeros in this formula than in Formula 1 because horsepower is a larger measure of power than watts. It takes 746 watts to equal one horsepower.

Horsepower is a useful measure of doing work. For this reason most electric motors are rated in horsepower rather than in watts. One horsepower is the equivalent of raising 33,000 pounds (15.000 kilograms) 1 foot (30.5 centimeters) in one minute. Table 3–3 gives the approximate horsepower of ideal windmill propeller blades for various wind speeds. This table helps to illustrate the two fundamental laws of wind power. Note that when the blade diameter is doubled, from 6 feet to 12 feet, the power is increased four times, as in the shaded squares in the 10 mph (16 kmph) row, from 0.12 to 0.47 horsepower. But, if the wind speed is doubled, the power is increased 8 times. This is illustrated by the shaded squares under the 12-foot (3.65-meter) blade column, where the power increases from 0.47 horsepower at a 10 mph (16 kmph) wind to 3.78 horsepower for a 20 mph (32 kmph) wind. Even a windmill as small as 6 feet in diameter can produce nearly a horsepower in a 20 mph (32 kmph) wind.

Table 3–3 Horsepower Produced by an Ideal Windmill Propeller

Wind Speed (mph) (kmph)	Blade Diameter (feet) (meters)			
	6 (1.82)	8 (2.42)	12 (3.65)	20 (6.09)
10 (16.04)	.12	.21	.47	1.32
15 (24.14)	.40	.71	1.60	4.95
20 (32.18)	.95	1.69	3.78	10.5
25 (40.23)	1.85	3.30	7.42	20.6

Conversion to Electrical Energy

If all of the 59.3 percent of the wind energy available could be extracted by a perfect windmill rotor and converted to electrical energy this amount would be less than expected. This is because additional losses in the conversion process further reduce the overall efficiency. In the step-up gear box and in the generator losses amount to approximately 30 percent (Fig. 3–4). Thus, the actual electrical power delivered by the wind generator is usually about 70 percent of the power calculated by using Formula 1 described earlier.

Fig. 3–4: Wind generator efficiency.

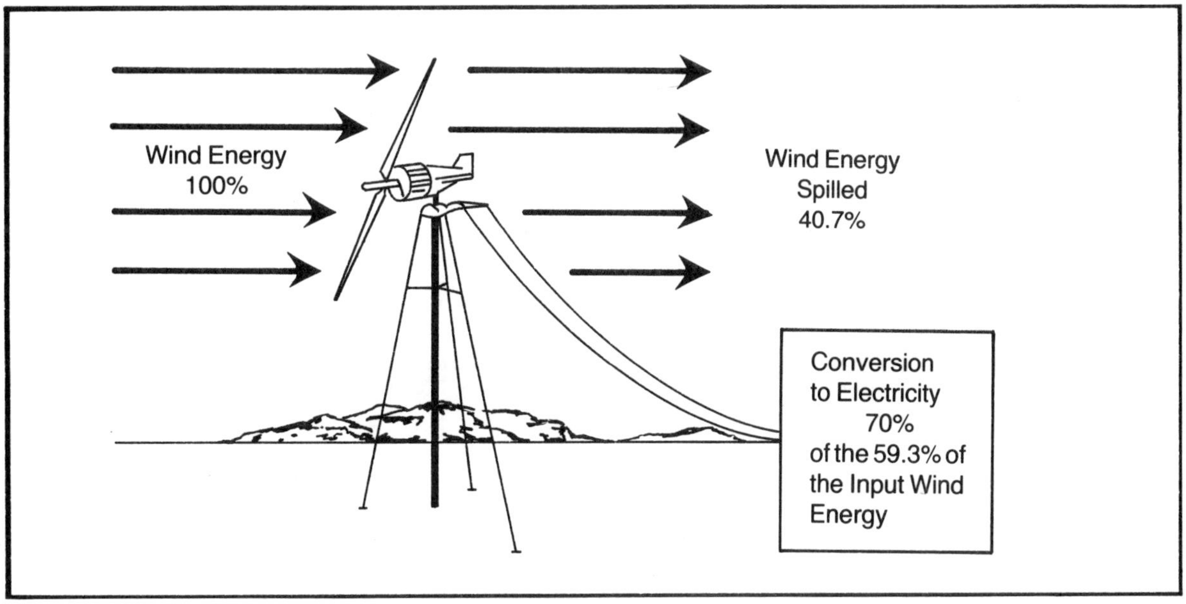

Table 3–4 incorporates this 70 percent efficiency for electrical conversion to calculations of wind generator power using Formula 1. In this table the power is given in kilowatts. One kilowatt is equal to 1000 watts.

Table 3–4. Kilowatts Produced by an Ideal Windmill Propeller Connected Through a Gearbox to at 70 Percent Efficient Generator

Wind Speed (mph) (kmph)	Blade Diameter (feet) (meters)			
	6 (1.82)	8 (2.43)	12 (3.65)	20 (6.09)
10 (16.04)	.062	.110	.298	.889
15 (24.14)	.029	.371	.837	2.325
20 (32.18)	.496	.882	1.984	5.910
25 (40.23)	.969	1.722	3.875	20.766

Note again that doubling the diameter of the blades increases the power in watts four times, while a doubling of the wind speed increases the power eight times.

Energy Production Estimation

Table 3–4 shows that a well-designed wind generator with a 13-foot propeller may produce nearly 2 kilowatts in a 20 mph (32 kmph) wind. This same machine may generate only one-fourth kilowatt, or 248 watts in a 10 mph (16 kmph) wind. If this wind generator is sold commercially as a 4kw machine, rated for a wind speed of 25 mph (40 kmph), how much power can it produce in a specific time period, such as one month?

The monthly (or weekly or yearly) energy production from a wind generator can only be estimated accurately if long-term wind survey information is available for the windmill site. Meteorological records, available for over 2000 locations in the United States from the National Climatic Center (Federal Building, Asheville, NC 28801) are an important first step, as are records from the nearest airfield or other weather station. The most valuable wind records, however, are those taken at the exact height and location of the wind generator. Even if these records are incomplete, they can be compared with the more extensive records from the nearest official recording station, and a correction factor can be included in the energy production estimates based on the meteorological records.

Table 3–5 shows a portion of a wind speed record from the National Climatic Center for the month of January, averaged over 7 years. This particular chart is called a Percentage Frequency of Surface Wind Direction and Speed. The entire chart shows the percentage of wind of a specific speed range distributed over the 16 compass points. Since the wind generator can turn, or yaw, to meet wind from any direction only the

SPEED (KNTS) DIR.	1–3	4–6	7–10	11–16	17–21	22–27	28–33	34–40	41–47	48–55	'56	%	MEAN WIND SPEED
N	.1	.3	1.3	1.1	1.0	.5	.2	.1	.8	.0	.1	4.8	15.8
NNE	.2	.1	.5	.7	.9	.8	.6	.1		.0	.0	3.8	19.8
NE	.2	.1	.6	.8	1.2	.8	.7	.1	.0	.1		4.5	19.4
ENE	.1	.1	.3	.8	1.3	1.5	1.2	1.0	.3	.2	.9	7.7	30.0
E	.2	.4	1.1	1.9	1.5	1.0	.9	1.0	1.0	.3	.2	9.5	23.7
ESE	.1	.1	.5	1.0	1.3	1.3	1.1	.8	.8	.7	.1	7.9	27.4
SE	.1	.3	.6	.7	1.1	1.6	1.3	.9	.4	.3	.1	7.6	25.1
SSE	.0	.2	.3	.2	.3	.4	.4	.7	.5	.8	.3	3.9	32.1
S	.3	.3	.6	.9	1.0	.7	.7	.8	.5	.4	.2	6.4	24.9
SSW	.1	.1	.5	.6	.4	.7	.7	.3	.0	.2	.2	3.7	24.9
SW	.1	.2	.6	1.2	.9	1.1	.9	.5	.3	.2	.2	6.3	23.9
WSW	.0	.2	.8	.9	1.1	1.1	.9	.6	.3	.2	.0	6.1	23.0
W	.2	.3	.7	.9	1.3	1.4	.9	.4	.1	.1		6.3	20.7
WNW	.2	.3	.5	1.1	1.0	1.1	.7	.6	.2	.1		6.0	21.3
NW	.3	.4	1.7	1.9	1.4	1.4	.7	.6	.2	.0		8.7	18.1
NNW	.0	.2	.8	1.7	1.5	1.1	.4	.2	.0	.1		6.0	18.6
VARBL													
CALM												.9	
	2.3	3.7	11.4	16.4	17.3	16.6	12.2	8.7	4.8	3.7	2.1	100.0	22.9
								TOTAL NUMBER OF OBSERVATIONS					22.9

Table 3–5. Percentage Frequency of Wind Direction and Speed from Hourly Observations (Amchitka Island, Aleutian Islands) for the Month of January 1944–50

total percentage of wind from all directions for a particular speed range is important.

The 4kw wind generator will produce very little power at winds of below 10 mph (16 kmph). It will generate no power at winds greater than 34 mph (58 kmph) because, as is true with many small wind generators, a protective mechanism turns the blades out of the wind at approximately this speed to protect the machine from possible blade breakage. In Table 3–6 the percentage at the useful wind speed ranges, from Table 3–5, are converted to hours per month:

Table 3–6. Energy Production Estimation

Wind Speed Range (mph) (kmph)	Hours per Month	Approximate Generator Output in Kilowatts	Monthly Output (kwh)
11–16 (17.7–25.75)	122	0.68	83
17–21 (27.35–33.79)	129	1.7	219
22–27 (35.4–43.45)	124	3.8	471
28–33 (45.06–53.1)	91	7.3	664
			Total 1,437 kwh

The third column shows the approximate wind generator output in kilowatts for each useful wind speed range. Each value is found by taking 70 percent of the power in watts calculated by using Formula 1. The answer in watts is then changed to kilowatts by dividing by 1000.

The fourth column shows the estimated monthly output of electrical energy for each useful wind speed range. This is found by multiplying column 1, the number of hours per month, by column 3, the output in kilowatts. The total estimated monthly output is found by adding the numbers in column 4. It must be remembered that even this energy estimate is only a rough guess of the monthly output at the exact location of the wind generator. The exact power generated may vary by as much as 50 percent. Only a careful study of the wind conditions at the site, including changes in wind speed with height, will give wind information sufficiently accurate for a realistic wind power production estimate.

WIND GENERATOR SELECTION

Some wind generators produce DC (direct current), while others supply AC (alternating current). In general, most of the smaller machines that would be purchased for an individual home, ranch, or farm generate DC, while the larger units, particularly those built under government-sponsored contracts or by power companies, produce AC. Wind generators that produce DC are usually connected to batteries that can store energy for use when the wind is calm, thus they are ideal for isolated locations not served by power company lines. As mentioned earlier in this section, DC cannot be stepped-up or stepped-down to higher or lower voltages by the use of simple transformers, as is done with alternating current, and is usually used at the same voltage that is generated. But, DC can be converted into AC by the use of modern transistor devices called inverters.

Wind generators that produce AC are usually larger machines and are connected to utility company power grids for distribution of the electricity they produce. This is because the voltage of alternating current can be changed easily by using transformers, and because AC cannot be stored as simply as DC. AC first must be changed to DC, or converted to another form of energy. Wind generators that produce AC usually serve areas already supplied by AC power from other types of generators. When the wind is strong, the other generating stations reduce production of electricity. When the wind weakens, the other stations increase their production. Some individuals in the United States with AC wind generators connected to their home wiring have made arrangements with the power company that permits them to send surplus wind power electricity back into the power lines, and they receive credit for this electricity. In some parts of the country, however, power companies do not encourage private individuals to connect an AC wind generator to the home wiring system.

Many years of experience with wind generators in all parts of the world generally indicates that DC wind generators are the most versatile for individual homesite applications. If power company AC lines are also present, there need not be a direct link between the two power sources. The DC wind generator installation is designed to supply the greatest portion of the power demand, and DC motors are utilized wherever possible, to operate directly from the wind generator, or from the battery storage system. Appliances that cannot be changed to operate from direct current are connected to the power company AC lines. Where AC power is not available, an electronic inverter, connected to the battery storage system, is used to supply alternating current to these appliances. Line loss and utilization problems connected with low-voltage DC wind generators are eliminated by selecting a wind generator with a 120-volt DC output and by installing a 120-volt battery storage system.

Types of Wind Generators

Wind generators are classified as either horizontal-axis or vertical-axis, depending on whether the rotor shaft is horizontal or vertical. The propeller wind generators and the multi-bladed water-pumping farm windmill are familiar examples of horizontal-axis wind machines.

Horizontal-axis wind machines may be single-, double-, three-, or four-bladed, or multi-bladed (Fig. 3–5). A single-bladed wind rotor requires a counter-weight to eliminate vibration, but this design is not practical where icing on the one blade could throw the machine out of balance. The two bladed propeller is the most widely used because it is strong and simple

Fig. 3–5: Horizontal-axis blade configurations.

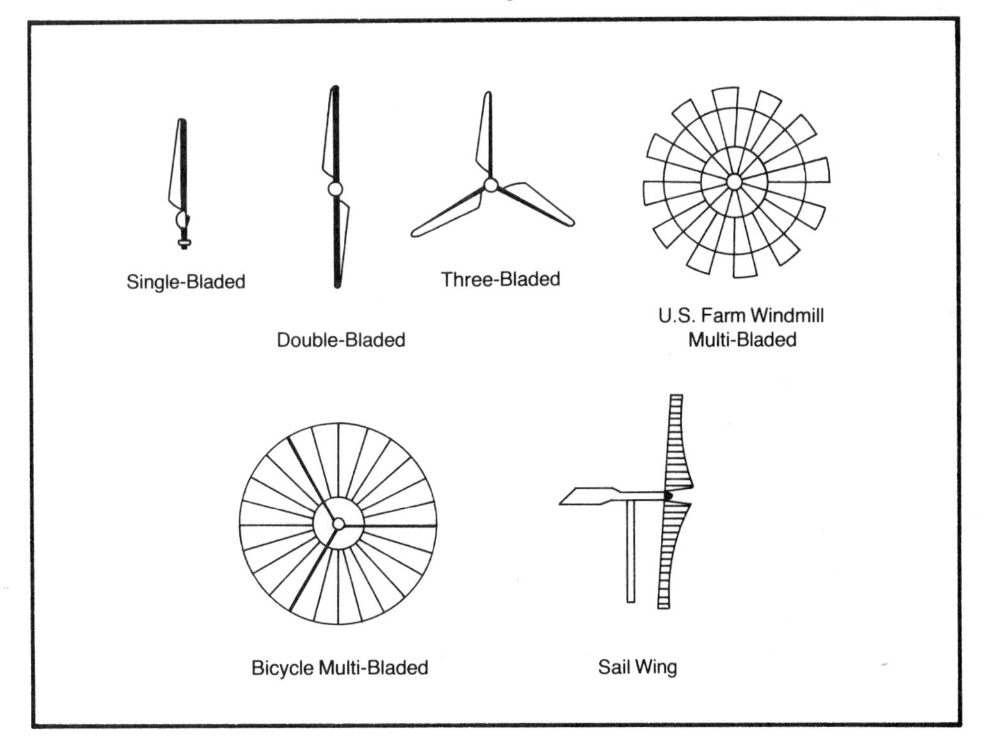

and less expensive than the three-bladed propeller. The three-blade propeller, however, distributes stresses more evenly when the machine turns, or yaws, during changes in wind direction. This was first pointed out by the pioneer of American wind generators, Marcellus Jacobs, who in 1927 settled on the three-bladed propeller for his machines. Briefly, he discovered that a two-bladed machine with a tail vane would yaw in a series of jerking motions because at the instant the propeller was vertical it offered no centrifugal force resistance to the horizontal movement of the tail vane in following changes in wind direction. At the instant the two-bladed propeller is in the horizontal position its centrifugal force, which is at maximum, resists the horizontal movement of the tail vane. The three-bladed propeller cures this problem, which produces a high level of vibration in the machine, by creating a steady centrifugal force against which the tail vane moves smoothly to shift the direction of the wind generator. Jacobs estimated that if his 15-foot (4.5-meter) propeller had two blades instead of three, 1100 pounds (500 kilograms) of gyroscopic resistance force would be applied and then removed to the movement of the tail vane twice during each revolution of the rotor.

Two relatively new horizontal-axis designs are the bicycle multi-bladed and the sail wing. The rotor of the bicycle WTG (wind turbine generator) is built much like a bicycle wheel in that the key structural members (narrow blades instead of spokes) are held in tension between the solid metal rim and the hub. This allows the rotor to be very light and strong. This design permits power to be taken from the rim by using a generator with a friction-drive hub.

The sail wing rotor is a modern descendant of the ancient jib-sail windmill; but, although made from metal tubing and cloth, it takes the general shape of the modern solid propeller. Sail wing blades use a metal tube to form the leading edge of the blade, and short bars from the tube form the tip and root of the blade. A cable is stretched between the tip and the root bars to serve as the trailing edge of each blade. The blade covering is made in the form of a cloth sleeve which slips over the blade frame.

Wind rotors for horizontal-axis machines are classified as either *upwind* or *downwind.* In upwind machines (Fig. 3–6) the blades are located ahead, or upwind, of the tower; in downwind machines they are located behind, or downwind of the tower. Small wind generators are usually of the upwind type for two principle reasons: (1) a simple tail vane is all that is needed to keep the blades pointed into the wind, and (2) a furling mechanism that turns the blades out of the wind stream to protect the machine from high winds is easier to design and fabricate for an upwind rotor.

The downwind configuration is usually preferred for larger machines, where a tail vane would not be practical. While the largest downwind machines are usually steered by a pilot wind vane coupled to a gear drive, as in the NASA Mod-O, OA, and 1 machines, smaller downwind machines have a natural tendency to yaw, or turn with changes in wind direction, and will automatically align with the wind.

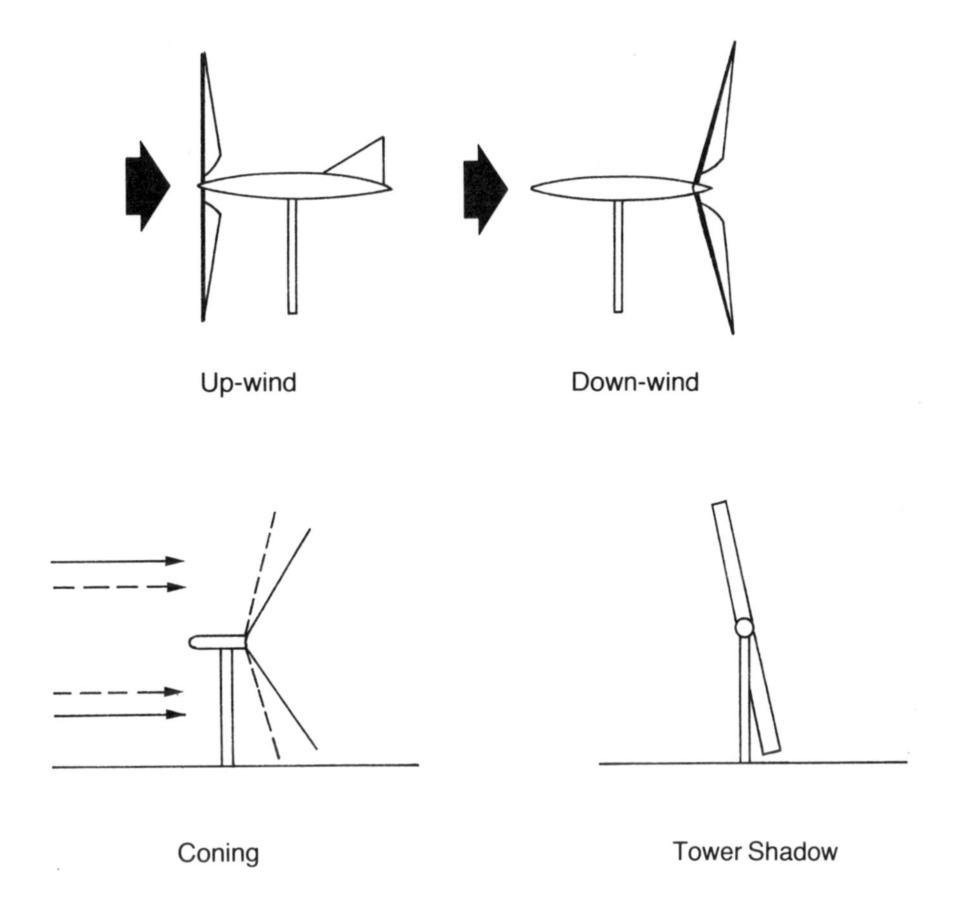

Fig. 3–6: Up-wind and down-wind designs.

Downwind machines are also preferred when the design calls for the blades to cone, or tilt downwind from the hub, where the blades are hinged. Coning is a very effective means of protecting large windmill blades from excessive stresses during high or gusty wind conditions. Coning is not practical in upwind machines since the blades would hit the tower structure. One problem with the downwind configuration, however, is tower shadow. The tower acts as a barrier to the windstream, and each time a rotating blade passes the tower it is subjected to changes in wind speed, which causes stresses that vary with the exact amount of wind blocked by the tower. NASA has had to modify the tower at the Mod-O site at Plum Brook to allow more wind to pass unhindered through the tower framework to reduce tower shadow stresses on the blades.

Vertical-axis wind generators have an important advantage over horizontal-axis machines in that they do not have to yaw, or turn when the wind changes direction, as they respond to wind from any direction. Although certain vertical-axis windmill designs are undoubtedly the oldest form of man-made machines for harnessing wind power, in recent times the vertical-axis concept was neglected until this century.

Two well known vertical-axis designs are the Savonius, or S-rotor, first patented in 1929, and the Darrieus, patented in 1931 (Fig. 3–7). Each design retains the name of its inventor. The S-rotor is predominantly a drag device, which means that its rotor tips do not travel faster than the wind speed. The split Savonius, however, does exhibit better performance since during part of each rotation it functions as a two-stage turbine. Savonius applied his rotor to water pumping and building ventilators. Its slow rotational speed and low power make its use for electrical generation impractical. The S-rotor may be thought of as the vertical-axis counterpart to the American multi-vane water-pumping farm windmill. Both are self-starting under load at low wind speeds and display high torque at low rotational speeds, attributes ideal for mechanical applications such as water-pumping.

The Darrieus and its derivatives, such as the Δ-Darrieus, are true lift-type machines with airfoil-shaped blades. They are the vertical-axis counterpart of propeller wind generators and have relatively low torque at slow wind speeds but develop good power as the speed increases. They are usually not self-starting and must be spun by an electric motor when the wind speed reaches acceptable levels. The giromill, from the words cyclo*giro* wind*mill,* is a recently developed vertical-axis wind machine with two or more sets of blades that flip, or change orientation, to the wind twice

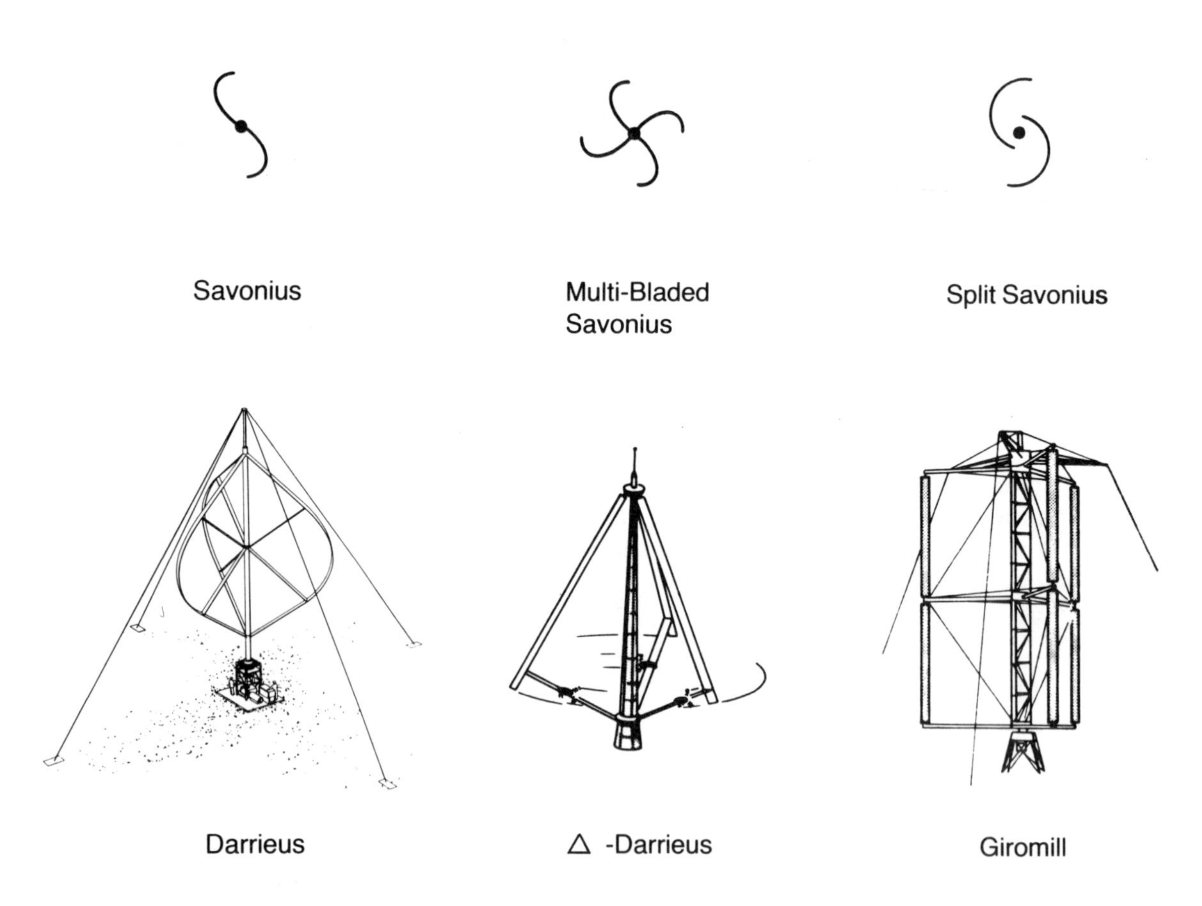

Fig. 3–7: New windmill blade designs.

ROTOR TYPE	Cp MAX	COMMENTS
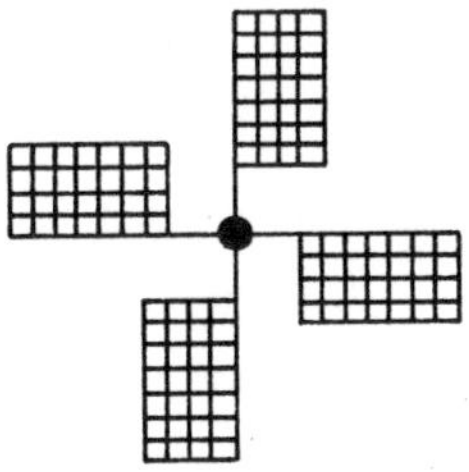 Dutch	0.17	High Torque, Low RPM Inefficient Blade Design
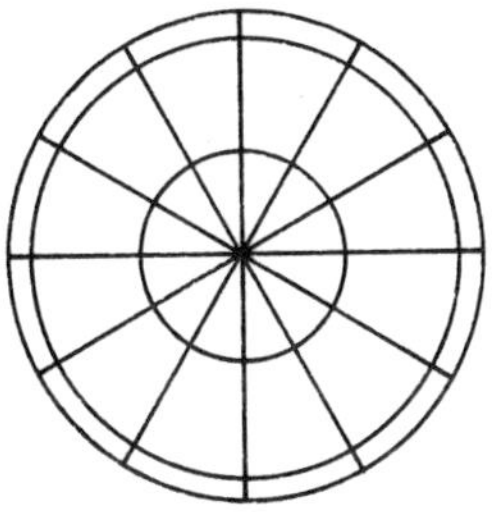 Farm	0.15	High Torque, Low RPM High Losses
 Modern Propeller	0.47	Low Torque, High RPM Efficient Blade Design

Fig. 3–8: Maximum power coefficients for various rotor designs.

during each revolution of the rotor. Symmetrical blades are used so that at each flip the other side of the blade is exposed to the driving force of the wind. Between the flip regions, about 90° to the wind direction, the blade angle is adjusted to keep the blades angled to the wind for maximum driving force. The giromill runs slower than a similar size Darrieus, but develops more power.

Wind Generator Efficiency

Windmill rotors with four to eight blades are generally limited to the colorful Dutch-type water-pumping and grain-milling units. The multi-bladed farm-type windmill rotor, often called a fan, is confined to applications such as water pumping because of its slow rotational speed. Both designs develop high torque at slow rotational speed, but have a low efficiency because their power coefficient, C_P, is small when compared to propeller-type rotors (Fig. 3–8). Propeller wind rotors and the lift-type vertical-axis rotors are the most efficient because the tips of the blades revolve at a high speed compared to the wind velocity. This is called the tip-speed

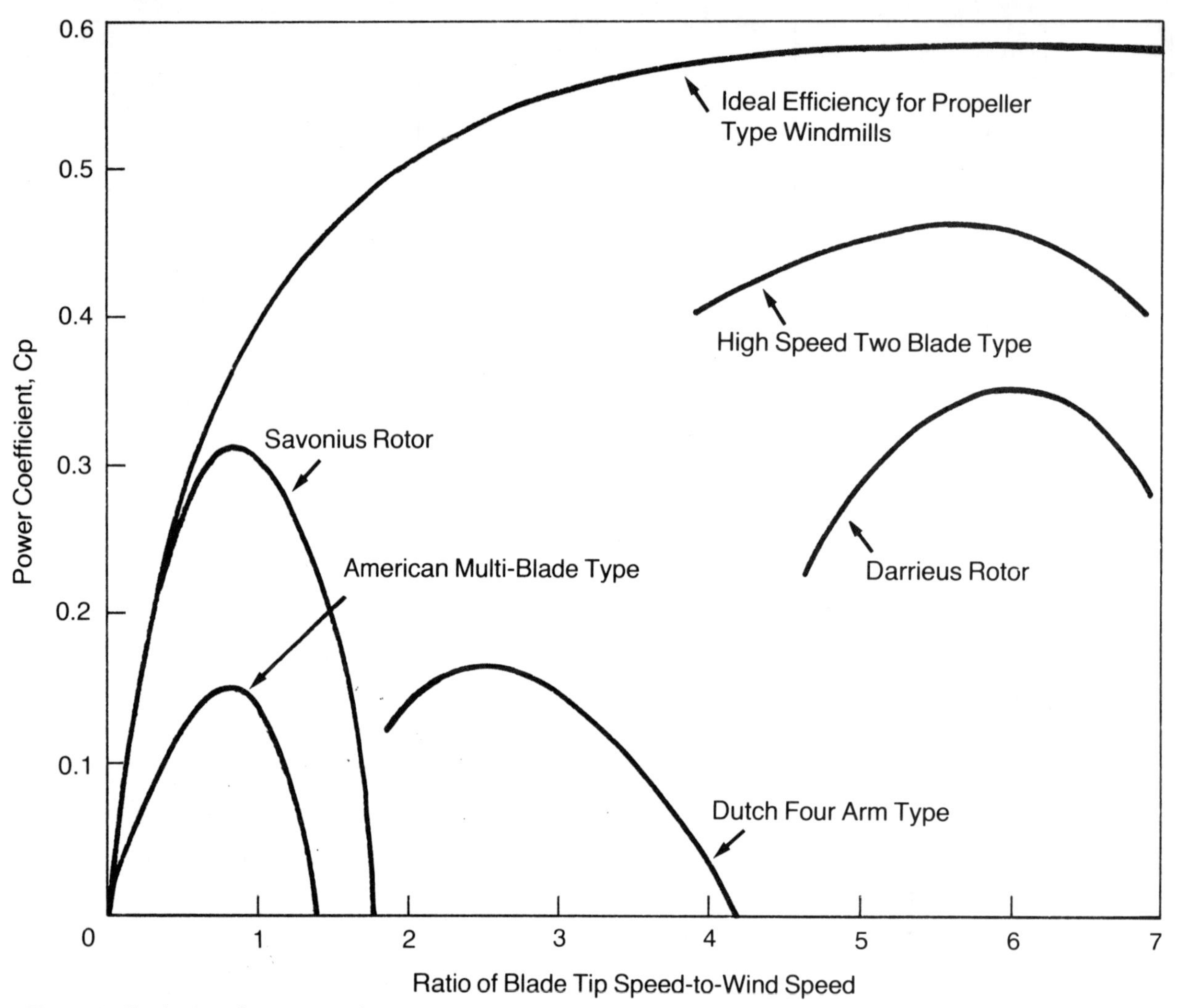

Fig. 3–9: Typical performances of wind machines

ratio, and is defined as the ratio of the speed of the blade tip to the speed of the wind.

The power coefficient of each wind rotor design is the power delivered by the rotor divided by the total power available in the wind that strikes the area swept by the rotor. The power coefficient of an ideal wind rotor varies with the tip-speed ratio, and approaches the maximum of 0.593, or 59.3 percent, of the total power available in the wind when the tip-speed ratio reaches a value of five or six. The relationship of the power coefficient to the blade tip-speed ratio for most wind rotor designs is shown in Fig. 3–9.

Each wind generator design, whether a commercially manufactured machine or home-built from plans or through personal inventiveness, must be evaluated on an individual basis. Some of the many factors that influence efficiency and power output under a variety of wind conditions are the number, shape, and size of the blades; the pitch, or angle, of the blades; the *cut-in* and *cut-out* wind speeds of the machine; and the type of governor or furling to protect the machine in high winds.

The cut-in and particularly the cut-out wind speed of each wind generator design has a pronounced effect on the power output of the machine. A typical power output curve for an AC wind generator is shown in Fig. 3–10. Most AC wind generators must operate at a constant rpm to keep the frequency of the alternating current at a constant value. The blades of the NASA Mod-O machine, for example, turn at 40 rpm, while the gearbox steps-up this speed to 1800 rpm, which is required by the synchronous alternator to produce 60-cycle current. At the cut-in wind speed the blades are turning at 40 rpm, and the machine begins to produce electricity. As the wind speed increases, the blade rpm is kept constant by increasing the load on the machine, that is, by producing more electricity. With increasing wind speed, the amount of electricity generated continues to increase. This region is the slope of the curve in Fig. 3–10. At the rated wind speed and power output, V_{RATED}, the blades are still turning at 40 rpm, and the power produced by the machine is at its design maximum. Should the wind speed continue to increase, the blade pitch angle is gradually changed to keep the blades turning at 40 rpm, and the frequency of the alternating current produced remains at 60 cycles per second. In this manner a constant power output, the flat portion of the graph, is maintained between the rated wind speed and the cut-out wind speed. At the cut-out wind speed, the blades are feathered, or turned edgewise to the wind, to protect the machine from high-wind damage. At

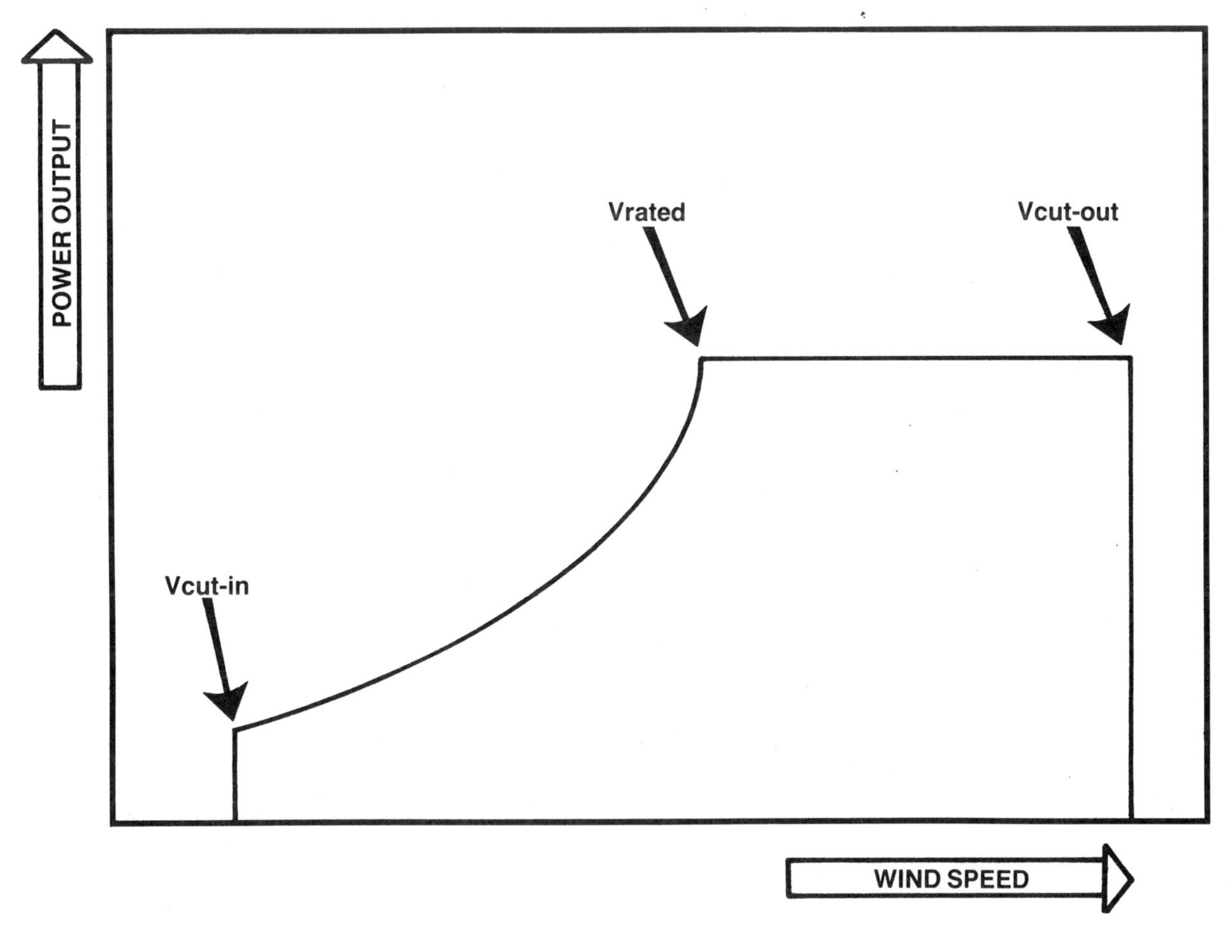

Fig. 3–10: Power output versus wind speed

cut-off, the alternator is disconnected from the power lines, and the power generated drops to zero. With this type of AC wind generator, a portion of the power available in the wind at speeds greater than the rated value is sacrificed to keep the frequency of the alternating current at a constant value. A well-designed AC wind generator should have its rated wind speed carefully chosen so that a minimum of wind power is not utilized in generating electricity. To accomplish this goal most medium and large AC wind generators have a rated wind speed specifically chosen for the wind conditions anticipated at the site where the machine is to be installed.

Most DC wind machines generate an increasing amount of direct current until the wind speed reaches the cut-out point. With some machines a flyball governor will then feather the blades to protect the machine from high winds, and the power output will drop to zero. Other machines incorporate a furling mechanism that swings the rotor out of the windstream to protect the blades and generator from damage. With this type of wind generator the power output at the cut-out wind speed also drops to zero. Still another type of DC wind generator utilizes an air brake governor with flaps that open in high winds to control the speed of the propellor (Fig. 3–11). Machines of this type can be designed to produce power even when the air brake is operating, thus producing a constant power output. Therefore, with DC wind generators the rated wind speed and power output point may be placed at any location on the power output curve.

PUBLIC ACCEPTANCE OF WIND GENERATOR DESIGNS

Another increasingly important aspect of harnessing wind power is the visual impact that one or more wind generators have on the surrounding area. While all wind machines are beautiful to windmill enthusiasts, other members of the public are not pleased with the thought of forests of windmills projecting into the sky. A recent study prepared for the National Science Foundation as part of the Federal Wind Energy Program administered by the Department of Energy sampled public reaction to various windmill designs sited in a variety of locations.

The technical feasibility of utilizing wind generators to produce electricity is well established. For some locations, such as isolated areas or small towns removed from major population centers, electricity could be generated more efficiently by wind machines than by other means. In more settled areas, wind machines could supplement existing electrical generating equipment.

Growing public interest in all aspects of the environment indicates that before wind generators can be implemented on a large scale, attention needs to be given to the positive or negative public reaction to wind machines. Numerous instances already exist of technological developments that were feasible and economic but which, because of public dissatisfaction, had to be either modified or discarded altogether.

The question of public acceptance is particularly relevant for wind generators which would have to be placed in large numbers at various loca-

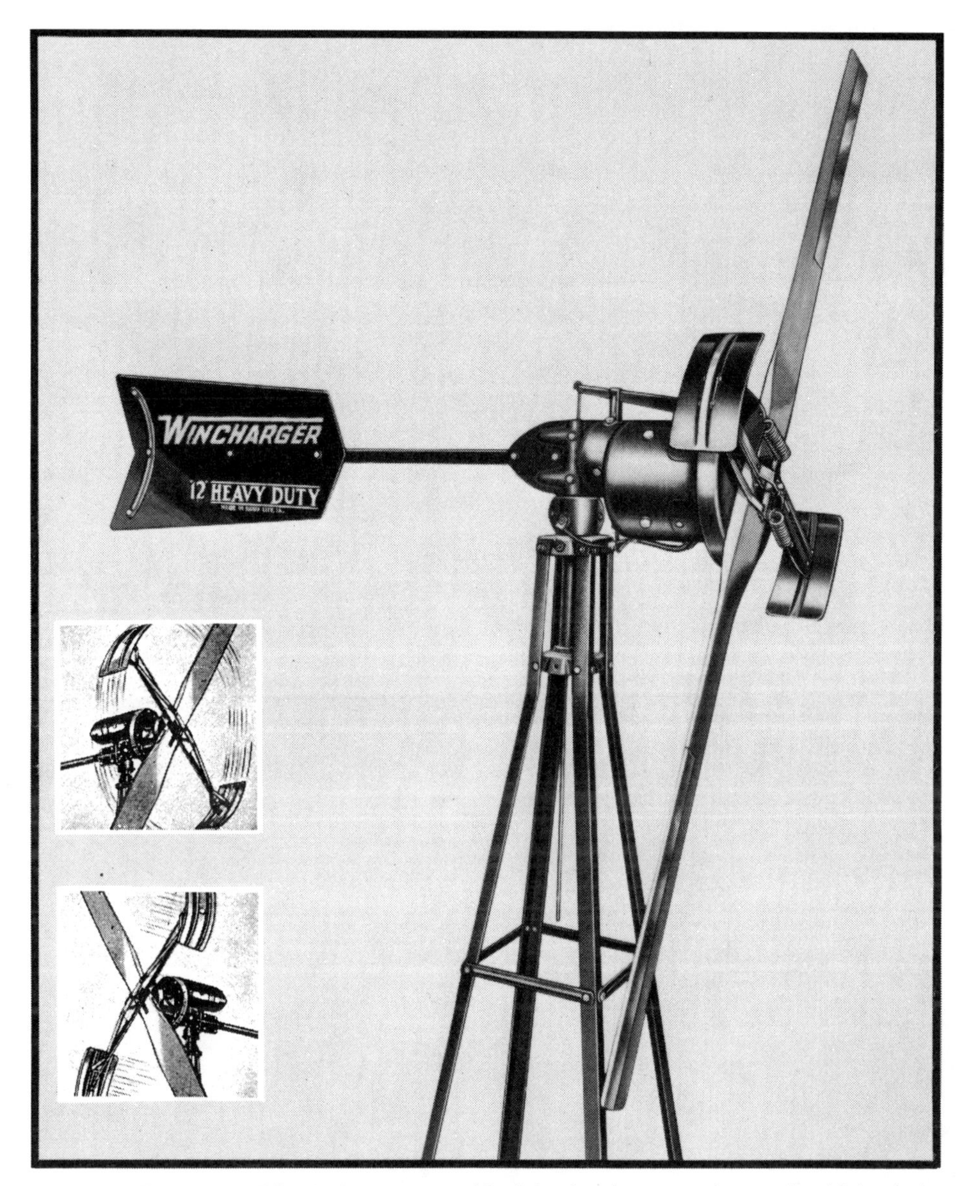

Fig. 3–11: DC wind generator with air brakes for protection against high winds. (Winco Division of Dyna Technology)

tions around the country and would occupy highly prominent positions, visible to all people nearby. The willingness of the public to accept such a change of the landscape is likely to be a major determinant in implementing large numbers of wind generators.

The subject of aesthetic value is only occasionally mentioned when wind generators are discussed. Most people know very little about wind power and even less about how a wind generator extracts energy from the wind. The public thinks of windmills only in terms of the Dutch tower-type or the multi-vane water-pumping farm type. While many people may have read one or two articles in newspapers or magazines about the use of wind machines for generating electricity, most people have virtually no

Demonstration on the Mall at the Washington Monument, Washington, D.C. for National Day—May 3, 1978

ROTOR BLADES FOR LARGE DIAMETER WIND TURBINE GENERATORS
Kaman, under a NASA contract, has designed and fabricated the largest rotor blade ever made. Half the length of a football field and weighing approximately 12 tons, the 150-foot-long all-composite blade is being tested by Kaman to evaluate existing methods of predicting the physical properties, strength and dynamic characteristics of fiberglass blades of this unprecedented size during the design process. The project is part of a Federal wind energy program sponsored by the Department of Energy (DOE).
Photo: Kaman Aerospace Corporation

conception of how a wind generator produces electricity from wind power.

The long-term goal of the federal wind energy program should be to educate the public about types of wind generators that are the most efficient and thus the most desirable. As a short-term goal it would prove useful to sample current public opinion as to types of wind machines that are the most pleasing to the viewer.

Recognizing the need for such information, a study was undertaken to explore reactions of the general public toward different types of wind machines. The primary objectives of the study were twofold:

1. To provide information on public acceptance of different types of wind machines in a variety of settings.
2. For this information to serve as a base for more intensive studies, now or in the future, to determine shifts in opinion.

The study was carried out by conducting personal interviews with random samples of adults in six different parts of the country, with primary emphasis on rural and smaller urban areas. At the same time, some preliminary coverage of major urban areas of the country was provided.

The locations selected were:

1. Western Michigan by the shores of Lake Michigan.
2. Southeastern Wyoming.
3. The far western part of the state of Washington.
4. The eastern part of Rhode Island.
5. The Chicago metropolitan area.
6. A beach area in the Gateway National Recreation Area at Sandy Hook, New Jersey.

These areas are highly diverse and include a partly rural and partly small-town area in the Midwest, a primarily rural ranching area in the Great Plains, a lightly settled rural nonfarm area in the Pacific Northwest, a fairly heavily settled area on the East Coast, one of the major metropolitan areas of the country, and a heavily visited area in a national park. All of these are places with relatively high wind power.

A special feature of the study was that at the last location interviews were conducted while the respondents actually viewed a wind generator. As part of the study, a 6-kilowatt, three-bladed horizontal axis machine on a 40-foot (12-meter) tower was constructed near the beach at Sandy Hook, New Jersey, in the Gateway National Recreation Area.

The adults interviewed were questioned on a number of topics, including their attitudes toward sources of energy, and their knowledge of wind machines. In addition, they were shown color slides of different wind machines in various settings and were asked to indicate their preferences from an aesthetic point of view. Approximately 300 people were interviewed at each location, and for comparative purposes, a second group was interviewed at the national park after the wind generator had been constructed.

Opinions of those interviewed to different types of wind generators were requested while showing color slides in portable lighted slide viewers. The slides were shown in a random manner, and included:

1. Old Dutch windmill in a shore setting and in a hill setting.
2. Horizontal-axis wind generator on an open lattice tower in a flat-land setting and in a shore setting.
3. Horizontal-axis wind generator on a columnar (post) tower in a hill setting and in a flat-land setting.
4. Horizontal-axis wind generator on an old Dutch tower in a shore setting and in a hill setting.
5. Darrieus wind generator in a flat-land setting and in a hill setting.
6. Giromill in a shore setting and in a flat-land setting.

The results of the interviews strongly suggest that the old Dutch windmill design dominates people's preferences. This type of design was preferred by a 4-to-1 margin over the horizontal-axis wind generator on an

2 KW Dunlite wind generator powering channel navigation lights at the entrance to Port Adelaide, Australia
Photo: A Division of Pye Industries Sales Pty. Ltd.

open lattice tower and by a 3-to-1 margin over the horizontal-axis columnar tower machine. Respondents judged the horizontal-axis wind generator on an open lattice tower and the Darrieus machine about equally, but when the horizontal-axis machine was placed on an old Dutch tower the preference shifted to nearly 2-to-1 in favor of the horizontal-axis machine.

Of the six designs shown, the results of the study indicate the following preferences: The old Dutch windmill design is the overwhelming favorite, followed, in descending order, by the horizontal-axis wind generator on an old Dutch tower, the horizontal axis machine on a columnar tower, a tie between the horizontal-axis wind generator on an open lattice tower and the Darrieus machine, and, lastly, the Giromill. For the last three designs, a majority of those interviewed felt that the setting in which the machines were pictured would be more pleasing if the wind machines were not there.

Attitudes toward different designs did not appear to vary much from one setting to another. There was some tendency for the old Dutch windmill to be preferred even more when in a shore setting, and there was also some inclination for a greater portion of those interviewed to prefer the horizontal-axis machine on an open lattice tower in a prairie, or flat-land setting. This may be a result of many people's familiarity with the farm-type water-pumping windmill on a steel lattice tower.

Interviewing people in the presence of a wind generator, the machine at Sandy Hook, appeared to have little effect on their attitudes toward wind energy, but did have a noticeable effect on reactions to different designs. Viewing the machine at Sandy Hook led those interviewed to record more negative reactions to that particular type of wind generator, and to prefer other types more frequently. The wind generator erected at Sandy Hook was not especially attractive. Perhaps wind machine designers should strive to improve the appearance of their designs as much as possible without affecting performance.

One possibility is to design towers of a simple columnar nature. Towers of this type received good acceptance, although not as good as the Dutch-type tower. Conceivably, it may also be possible to design a modification of the Dutch-type tower that would be more simple and modernistic. People seem to be impressed favorably by modernistic designs, and if one could be incorporated with a Dutch-type motif, public reaction might be highly favorable.

Chapter Four

ENERGY STORAGE

One problem with generating electricity is a matter of too much or too little. It is a rare occasion when the amount of electricity being generated exactly matches the amount being used. The result is the need for storage, but storage of energy, especially electrical energy, is a big problem. And in spite of all the time that has passed since Ben Franklin flew a kite in a thunderstorm, no one has figured out a better way to store electricity than in storage batteries. Unfortunately, storage batteries are only good for holding direct current, DC, the kind found in an automobile. No one has come up with a way to store alternating current, AC, the kind used in most houses and businesses, even for a fraction of a second!

The most practical alternative developed thus far is conversion, that is, changing the energy into some form that later can be converted back into electricity. In this chapter we will describe it and other methods of converting wind energy into readily usable forms of electricity.

WATER STORAGE

In some parts of the country it is practical to store large amounts of water behind a dam. In other places, where this won't work, water can be pumped up to a tower or other storage system that holds the water until it is needed for conversion to electricity. When power is needed, the water is allowed to fall and turns a device called an impeller. The impeller is connected to a generator, which produces electricity. The cost of this type of water storage system will depend upon local conditions and available land space on which a relatively large body of water can be stored. But this kind of system cannot be used to respond quickly to short-term needs. A wind energy unit, complete with storage batteries, can be used to satisfy these kinds of rapid, short-term fluctuations in demand.

PNEUMATIC

Compressed air can be stored as an energy potential, and as the air is released, it can be used to drive a generator turbine. Research done on this method has shown it to be somewhat practical for large utility-size systems where the air is stored in caves, caverns, and abandoned salt mines. This is a low-efficiency type system, but it does work.

100' self-supporting Rohn Tower with Jacob's 2400 watt electric generator installed at Unarco-Rohn plant in Peoria, IL.

INERTIAL ENERGY

The uses of the inertial energy stored in a fast-revolving flywheel have been studied extensively. Inertial energy ultimately represents good efficiency and perhaps, at some future time, reasonable cost. One of the biggest problems in using inertial energy systems involves safety. To generate a sufficient amount of electricity, the flywheel must be rather big and must move very fast. The problem is that at extreme speeds, a flywheel will fail—that is, break. And when that happens, the pieces of the flywheel come off in large chunks that act like powerful missiles that can kill a human being. For safety reasons, these missiles must be contained. This problem is serious enough to make the entire system impractical.

Several attempts have been made to develop alternatives. British researchers have developed a flywheel made of banded rods instead of the usual heavy steel flywheel. This flywheel works the same as the conventional kind, except that when it fails, the pieces of the rods are small and relatively lightweight. A substitute for steel has also been sought. Materials that actually turn to dust upon failure and thus can be readily contained have been developed. Researchers feel that a flywheel fabricated from such materials may represent a worthwhile method of storing energy from a small wind system. The flywheel shaft can be connected to a suitable gear box to directly drive an electric generator. In such a system the flywheel is turned by energy provided by the wind-driven generator through a mechanical connection or by the wind system itself.

While there are many schemes that would seem suitable for storing energy developed from a wind generator system, the familiar lead acid storage battery is the most practical. This was the conclusion reached in "Applied Research on Energy and Conversion for Photovoltaic and Wind Energy Systems," a study prepared for the National Science Foundation and the U.S. Department of Energy. The conclusions of the study were: "While most of the energy storage concepts examined can be interfaced with wind or photovoltaic conversion systems in some manner, battery storage is the most universally attractive method for near-term (1985) use at all application levels. . . . Energy storage, in the future, is so heavily dependent on achievement of projected (lower) cost goals that it would be premature to rule out continued development effort of longer term candidate concepts."

While many years have passed since the discovery of the storage battery, it is still the most promising for storage of electrical energy developed by a wind energy system. It is important to realize that all schemes have been put forward with the assumption that local utility power is not available; that is, these schemes have concerned totally isolated systems that did not depend on other local generating sources, such as a utility company. However, in recent years many wind energy systems have been installed in areas where public utility power is available. The wind system in these cases has been used as a supplement to the existing power source. Such systems use the power generated by the wind system when it is available but use the utility source when the wind is not generating power. This type of system can feed excess electricity back to the local utility company when it is generating more power than is currently being consumed. In some parts of the country the local power companies have been cooperating with owners of wind energy systems, and this has led to a general acceptance of the concept. This means that when the user has a demand greater than what is being generated by the wind system, the system automaticaly draws from the utility company's power line. This eliminates the need for battery storage and reduces costs for the inverter system which is needed to produce alternating current from the direct current from the batteries. As a safety precaution for power company workers this system ceases to function when it no longer receives power from the utility. This is done to avoid the possibility that the wind system would send electricity back to the power company over a line thought to be down. Thus, when a utility power line is down, the wind energy system is also down until repairs have been made.

Of all the storage schemes discussed or envisioned to date the lead acid storage battery has proven to be the most economical and reliable. The materials used to make batteries have improved over the years, and today's highly reliable lead acid batteries have a greatly extended life and are also able to "deep cycle" many more times than older designs. The use of a combination of lead and antimony has increased the discharge cycle life to over 2000 discharge cycles. Battery life can last 7 to 8 years or more if they are properly maintained. Some lead antimony batteries have lasted 15 years.

UNDERSTANDING THE LEAD ACID BATTERY

Even though batteries of all types are used by nearly everyone, little is understood about their nature and abilities. Today's car battery requires little or no maintenance to keep it functioning well. Even the use of water in the electrolyte in some batteries has been eliminated. This is possible because in normal operation (easy starting) the battery rarely gets discharged. It starts the car and then is under a high rate of charge until it is brought back to the level of charge it had before it turned the starter. If you have ever had a completely discharged battery, you will recall that it took quite some time to get it back to normal. If this happens more than once or twice, you run the risk of ruining the battery—of running it down to where it won't be able to hold a charge. If this ever happened to you, it probably was in cold weather, when the motor doesn't turn over very fast because the oil in the engine is very stiff and the amount of energy required from the battery is substantially greater. Most people don't understand that the electrolyte in the battery also gets heavy and stiff and that this severely reduces its ability to put out the needed energy. Actual tests made while starting cars at 0°F (−18°C) show that the engine requires 250 percent more power than at 80°F (27°C) and that the battery has 61 percent or less of the power that it normally has at 80°F (27°C). The loss of power at low temperatures becomes very important when car-type batteries are used for primary storage in a wind energy system. Planning for this kind of loss is essential if the batteries are to be in an environment that reaches 40° (4.4°C) or lower.

Because the lead acid battery plays such an important role in wind energy systems, we will take a much closer look at its limitations and performance characteristics.

THE LEAD ANTIMONY (LEAD ACID) BATTERY

The lead acid storage battery is an electrochemical device that converts chemical action into electrical energy. It does not actually store electricity. It stores a *potential* to deliver electrical energy. Active materials within the battery react chemically to produce a flow of direct current when a current-consuming device is connected to the output terminals of the battery. Current is produced by a chemical reaction between the active elements of the plates and the sulfuric acid of the electrolyte.

The simplest unit of a lead acid storage battery is two plates, one positive and one negative, made of unlike material kept apart by a porous separator. This assembly is called an *element*. When this element is enclosed in a case and emersed in a sulfuric acid solution, it becomes a single *cell*. The voltage obtained from the simple cell is 2 volts of direct current (2 VDC). This small amount of current is suitable only for small electrical loads. The electrical current and power of the battery may be increased by combining several positive plates and several negative plates to form a multiple-plate cell, still having a total voltage of 2 VDC. From a theoretical viewpoint we could continue to combine more and more plates, make them larger and larger, and get more and more stored energy, though there are, of course, practical limitations, such as weight and bulk.

The important fact here is that the battery storage capacity has a definite relationship to the size, thickness, and number of plates. The number of plates used in an individual cell will depend upon the application of the battery. Typically batteries for starting diesel engines need to put out, or give up, very high current for a short period of time. These batteries have more plates (total surface area in contact with the electrolyte) than a battery that must deliver a more moderate amount of current for a longer period of time. Generally the greater the number of plates, the greater the current that can be drawn over a short time period. Batteries that have fewer, thicker plates cannot give up as much current, but they can provide moderate amounts of current for a longer period.

With these facts in mind, it becomes obvious that you must determine the work that the battery must do and how long it must continue to do this work. How do you define the work that a battery must do? Let's review the math and terms that must be thoroughly understood in order to manipulate the various measurements with which we must deal in order to select correct components and integrate them into a complete wind energy system.

ELECTRICAL ENERGY

In the world of electrical energy, work is measured in *watts.* Watts can be directly converted to *horsepower*. Horsepower is a measure of work done in a period of time: i.e., 746 watts = 1 hp. It takes 746 watts to equal 1 hp, assuming everything is 100 percent efficient, which, unfortunately, it isn't. If you want to run a 1 hp motor that is actually doing 1 hp worth of work, you must provide 746 watts of electrical energy *plus* additional energy to compensate for the inefficiency of the system (including the motor). If the efficiency of the motor and wiring system is 80 percent, then it would take 932 watts to run the motor (746 ÷ .80 = 932).

Voltage

Voltage is *electromotive force* (EMF). EMF is the difference in potential between the positive and negative elements of a battery—the available force to move electrons through a conductor. Voltage is often compared to the water pressure in a home plumbing system.

Amperes

Amperage is the measure of the flow of electrons (current) moving from the negative terminal to the positive terminal in a given amount of time. One ampere of current flows in a circuit when 6,240,000,000,000,000,000 electrons flow out of the negative terminal through a conductor and back into a positive terminal in one second! This can also be compared to the water flow in a home plumbing system if you compare voltage with water pressure and amperage with the actual flow, or amount of water, in motion.

Resistance

As electrons (electric current) flow through a conductor, there is a natural resistance to the flow. Some materials, such as copper, gold, and silver, have very low resistance. These materials are therefore known as good conductors since electrons flow easily through them. Resistance is influenced by different factors, such as temperature and cross-section diameter and length of the conductor. (Copper, our most frequently used conductor, increases its resistance as the temperature increases.) The resistance of a conductor is directly proportional to its length. The longer the conductor, the greater the resistance. A conductor with a thick cross-section (heavy gauge) has less resistance than a conductor with a small cross-section. Therefore, the resistance of a conductor is inversely proportional to the area of its cross-section.

Watts

As we have stated earlier, a watt is comparable to horsepower. It measures work done over a period of time. The dictionary defines the watt as "The unit of electrical power equal to one *joule* per second or a current of one ampere under an electrical pressure of one volt." If 1V at 1A = 1W (watt), then 10V at 10A = 100W. Just remember that $V \times A \times W$ (Volts × Amps = Watts).

Now that we have a basic understanding of some of the language of electricity, let's look at the relationships so that we can better understand the next steps. The math of basic electricity is simple and is displayed in the following formulas. First, we know that $V \times A = W$. If that is true the rest should be easy.

$$\text{If } V \times A = W$$
$$\text{then } V = W/A \text{ and also } A = W/V$$

Let's try an example. If your home uses 1200 watts of power and the voltage is 120V, then the current, or amps, is: 1200 W ÷ 120V = 10 amps. As we proceed, you will see the importance of this relationship as it applies to batteries and inverters. (It is good to review these formulas as you read other sections of this book.)

BATTERY STORAGE CAPACITY

As we mentioned earlier, all batteries have a built-in ability to give up electrical energy through an internal chemical reaction. The amount of energy is most often measured in terms of an *amp-hour.* Amp-hour (AH) simply means amps given up over a specified period of time. A 1AH battery that has a 1-hour rating will give up 1 amp for 1 hour, and an 80AH battery with a 1-hour rating will give up 80 amps for 1 hour. Industrial batteries and automotive batteries are rated at either the 8-hour level or the 20-hour level. Thus an 80AH battery rated at the 20-hour rate will give up 4 amps for 20 hours (80A ÷ 20 hours = 4 amps). The AH rating of any battery is good *only* for a specific set of circumstances. Thus, since industry standards generally indicate a temperature range of from 77 to 80°F (25 to 27°C), the AH capacity is only accurate within that temperature range. As the temperature drops, the electrolyte in the battery thickens and loses some of its ability to complete the chemical reactions necessary to make electricity. The colder it gets, the worse this condition becomes. Finally, the battery freezes. Look at Fig. 4–1, which shows graphically what happens to the AH storage capacity as the temperature drops from 77°F (25°C) to below the freezing level −22°F (−30°C).

If you study the chart closely, it is obvious that as the discharge rate increases (higher current is drawn from the battery), the AH (amp-hour) storage capacity is also reduced. This means that the method of rating batteries in terms of AH is only valid at one discharge rate and at one temperature. If either of these conditions changes, so does the storage capacity of the battery. These factors can be manipulated by using a curve like the one shown in Fig. 4–1. Nearly all battery manufacturers have such operating curves available for their products.

The relationship of temperature and discharge rate is very important to the potential owner of a wind energy system. Unfortunately, it is an important factor frequently overlooked—even by engineers—in the planning stages of a wind energy system. When a wind energy system is planned for a climate where the temperature is always below 32°F (0°C), lead acid

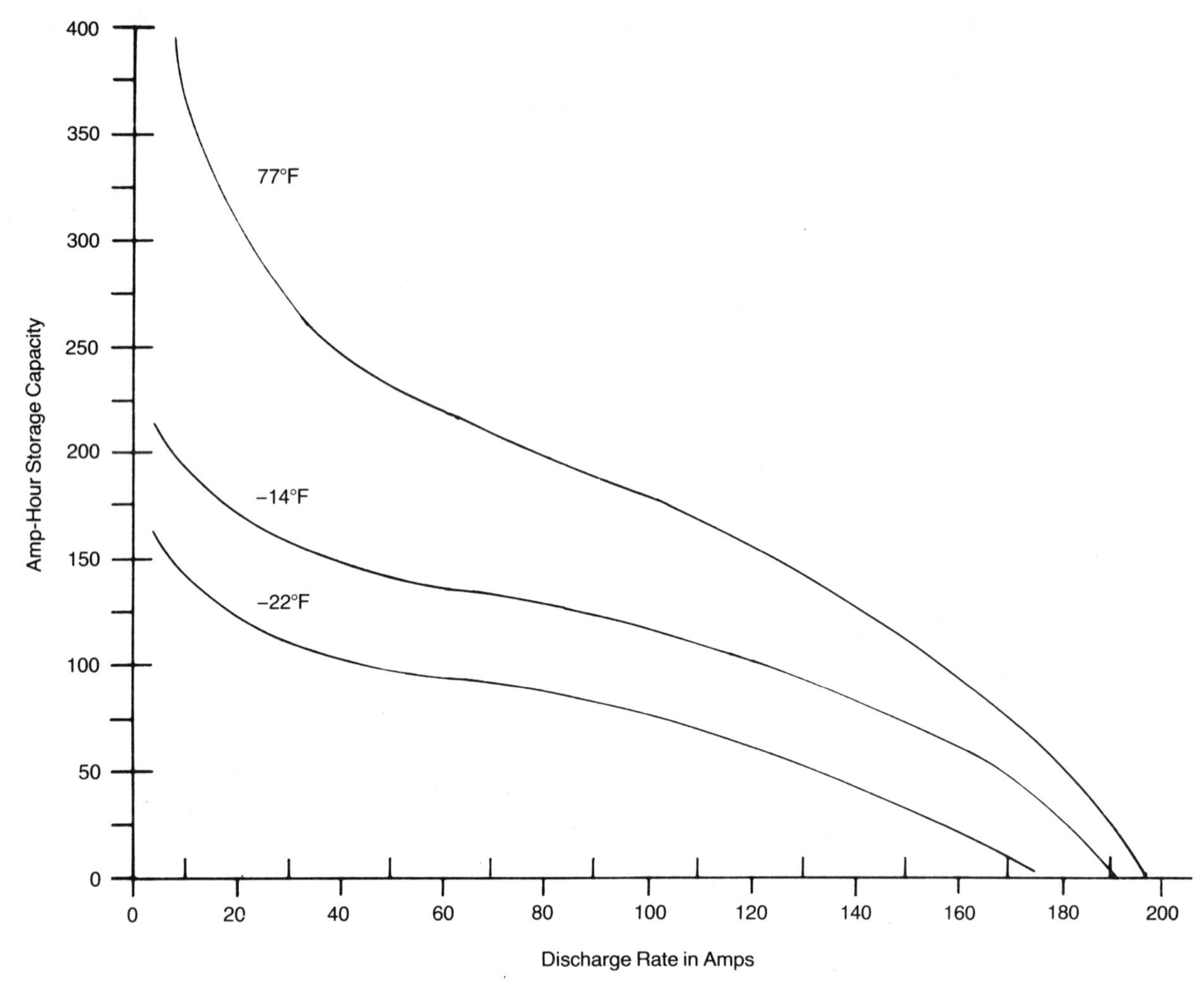

Fig. 4–1: Effect of temperature and discharge rate on amp-hour storage capacity

batteries become a marginal selection for storage purposes. However, nickel-cadmium (Ni-Cad) batteries perform very well in extremely low temperatures and do not suffer from the same limitations. We will discuss Ni-Cad batteries in more detail later.

Wind-driven electric generators and complementary systems produced today come in a wide and confusing variety of sizes and voltage outputs. Which system will be most suitable for you? If lead acid batteries are to be used as the means for storing energy, it might be useful to see how the electrical load will affect the choice of wind machine voltage output and, consequently, the proper battery configuration. Make a survey of the electrical usage in your own home. Since the basic measure of electrical energy used in homes is stated in *watts*, let's see just how many watts you use in your home during an average day. First, list all the kinds of electrical equipment that will be used and how long they will be in use. When this is complete, look at Fig. 4–2 to determine the wattage requirement for each item. The total is called the *peak,* or maximum, load point. This is the maximum amount of wattage that can be used at one time. For example, let's say that the following appliances are in your home. Your survey would look like this:

APPLIANCE	WATTAGE	APPLIANCE	WATTAGE
clock	2	hair dryer	381
coffee maker	894	refrigerator	321
attic fan	370	color TV	332
freezer	340	light bulbs 6 @ 60W	360
		TOTAL PEAK WATTS	3000

Once you know how many watts you will use, you can determine the amps (current) to be used when the voltage is 120 VAC. Earlier we discussed the formula watts divided by volts will give us amps ($W \div V = A$); therefore, in our example, 3000W ÷ 120V = 25A. This doesn't sound like much, but remember that it takes 1 horsepower to equal 746 watts at 100 percent efficiency. So your wind machine must deliver 4.02 horsepower to reach a 3000-watt output (3000 ÷ 746 = 4.02). When you consider all of the inefficiences of a system, you will find that it will take closer to 6 or 6.8 hp to deliver the 3000 watts you need.

But many of the earlier wind-powered generators used 36-volt batteries. What difference does this make? Plug the figures into the equation $W \div V + A$ and you'll see that it means quite a lot: 3000W ÷ 36V = 83.3A. This 83.3 amps is the current that the system must produce, and as we said earlier, current (resistance) is what drains a battery. Thus, the 120-volt system will run longer because it draws less current from the battery. The following table shows this inverse relationship between volts and amps.

3000	watts	at	120 volts	=	25.0	amps
3000	watts	at	48 volts	=	62.5	amps
3000	watts	at	36 volts	=	83.3	amps
3000	watts	at	24 volts	=	125.0	amps
3000	watts	at	12 volts	=	250.0	amps

There is another factor to be aware of: there are more batteries in a 120-volt system. Since all lead acid batteries have 2 volts per cell, you would need 18 cells for the 36-volt system and 60 for the 120-volt system. This represents a considerable cost difference, but in cases where 3000 watts are to be used in the home, the 120-volt system is a better choice for the wind generator.

Another fact that must be considered in a wind system is the effect that electric motors have on inverters, the batteries, and on the potential storage complex. We previously used a figure of 321 watts for the refrigerator (Fig. 4–2) which works out to be 2.68 amps at 120 volts. This is the actual power needed to run just the motor. However, it will take from three to six times this amount of current to get the motor started. This is 8 to 16 amps! For the freezer this means another 8.5 to 17 amps. If they start simul-

Appliance	Wattage	Appliance	Wattage
Air conditioner (window)	1,566	Range	8,200
Blanket	177	Range with oven	12,200
Blender	350	Refrigerator (12 cu. ft.) frostless	321
Broiler	1,436	Refrigerator (14 cu. ft.) frostless	615
Carving knife	92	Roaster	1,333
Clock	2	Sandwich grill	1,161
Clothes dryer (electric)	4,856	Sewing machine	11
Clothes dryer (gas)	325	Shaver	250
Coffee maker	894	Sun lamp	279
Diswasher	1,200	Television, b & w	237
Egg cooker	516	Television, color	332
Fan (attic)	370	Toaster	1,145
Fan (window)	200	Tooth brush	7
Freezer (15 cu. ft.)	340	Trash compactor	400
Freezer (15 cu. ft) frostless	440	Vacuum cleaner	630
Frying pan	1,196	Vibrator	40
Hair dryer	381	Waffle iron	1,116
Heater (portable)	1,322	Washing machine (auto)	512
Hot plate	1,257	Washing machine (non-auto)	275
Iron	1,088	Waste disposer	445
Light bulbs (six 60-watt)	360	Water heater	4,474
Oven, microwave	1,500		
Oven, self-cleaning	4,800		
Radio	71		

Fig. 4–2: Appliance/wattage chart.

taneously, you could overload the 3000-watt, 25-amp system. If the starting surge did not last too long (which it normally doesn't), everything will be under control unless all of the other equipment in the house is being used and if they started at the same time. Fortunately, the odds against any two units starting at exactly the same time are very high—perhaps 100,000 to 1.

What is or should be obvious at this point is that a wind energy system will require that the *owner consistently manage* his system and be fully aware of its limitations. Unlike the power available from the public utility, wind energy power is limited. A public utility is staffed by fully trained personnel. You need to be prepared for system failure as if you were the local power company, which, in a sense, you are. This doesn't mean that you have to be an electrical engineer, but it does mean that if you cannot handle the potential problems yourself, it is most important that you have someone available who can. A good local electrician or, perhaps, the manufacturer or the distributor from whom you buy the equipment may have competent service people available. But the truly successful systems are owned either by "handy folks" who understand and enjoy working with mechanical and electrical devices or by those who have ready access to an experienced equipment dealer.

LEAD CALCIUM BATTERIES

We have discussed the merits of the lead antimony (lead acid) storage battery. During the last few years there has been much talk—and a great deal of promotion—regarding the more expensive lead calcium type of battery. The principle advantage of this battery is its long life and low maintenance. This is particularly true when the lead calcium battery is used in stand-by power systems for emergency lighting, computers, communication links, etc., because it holds itself at a full charge for very long periods of time with no more than a *float*, or maintenance charging, level. In a wind energy system, however, the batteries are usually at work discharging or being charged and as a result go through a full discharge cycle at regular intervals. Under these conditions it is important that the battery system be designed to *deep-cycle discharge* and not sit unused for long periods of time. Since the lead antimony battery is designed to do this, and the lead calcium is not, the cycle life of the lead antimony battery is as much as 100 times greater than that of the lead calcium battery. If lead calcium batteries are used in a wind energy system, they do not have longer life and lower maintenance than lead antimony batteries.

Fig. 4–3: Deep-cycle battery with catalytic caps. (Delatron Systems Corporation)

The use of catalytic caps on lead antimony batteries (Fig. 4–3) will markedly reduce the number of times that the battery bank requires the addition of water to the cells. The catalytic caps recombine the hydrogen and the oxygen released during the charge cycle and return it to the battery as water. Although the efficiency of the conversion is not 100 percent, it does extend battery watering periods up to 1 year. And even during heavy use, the battery watering periods can be extended to 6 months. The most recent quote for catalytic caps is about $5 per cell, or $300 for a 60-cell 120VDC battery bank. In addition to water conservation, the caps also have a worthwhile safety feature. Very explosive hydrogen gas that is generated within the battery is isolated from the air outside the battery. This eliminates the possibility that a strong spark could ignite this gas.

ELECTRIC GENERATOR SETS, MOTOR DRIVEN

In a wind energy system it is quite possible to use all of the energy stored in the batteries and the wind and still not generate any useful power. When this condition arises, it is advisable to have some form of backup power to recharge the batteries or to generate the power needed during this calm period. Many owners have solved this problem by adding a gasoline, diesel, or propane engine-driven electric generator. Some environmentalists have used wood-burning steam engines for generating power. Since the generator usually will deliver more power than is needed for immediate use, it is important to try to load down the generator to its full capacity in an effort to conserve fuel. This is frequently accomplished by having the generator, operating at full capacity, charge the batteries to their full capacity, shut down the generator, and begin using the stored energy in the battery bank. This function also can be automated so that electric power is always available. An automated system monitors the condition of the battery bank, automatically starts the generator, when the batteries are low, and shuts off when the generator has fully recharged the battery bank. Alarms which sound when the generator fails to work can be added.

The possibility of being out of power clearly points up the importance of being able to manage your system by yourself or with expert technical assistance and the usefulness of automation. Some degree of automation is practical (see Fig. 4–4), but it costs money. The more automatic the system, the higher the cost. Completely automated wind systems can be found in radio, television, and other communications links which remain totally unattended in remote areas for as long as 12 months. Thus, although cost may be high, you will have a more dependable system.

Many of these systems are not only located in remote areas, but they are also at high elevations. The elevations of the tower sites is a definite benefit since winds tend to be more powerful and consistent at higher levels. There are cases on record where the backup generator has been called into service only once or twice in a single year. The practicality of wind energy for this kind of application is obvious since bringing power to

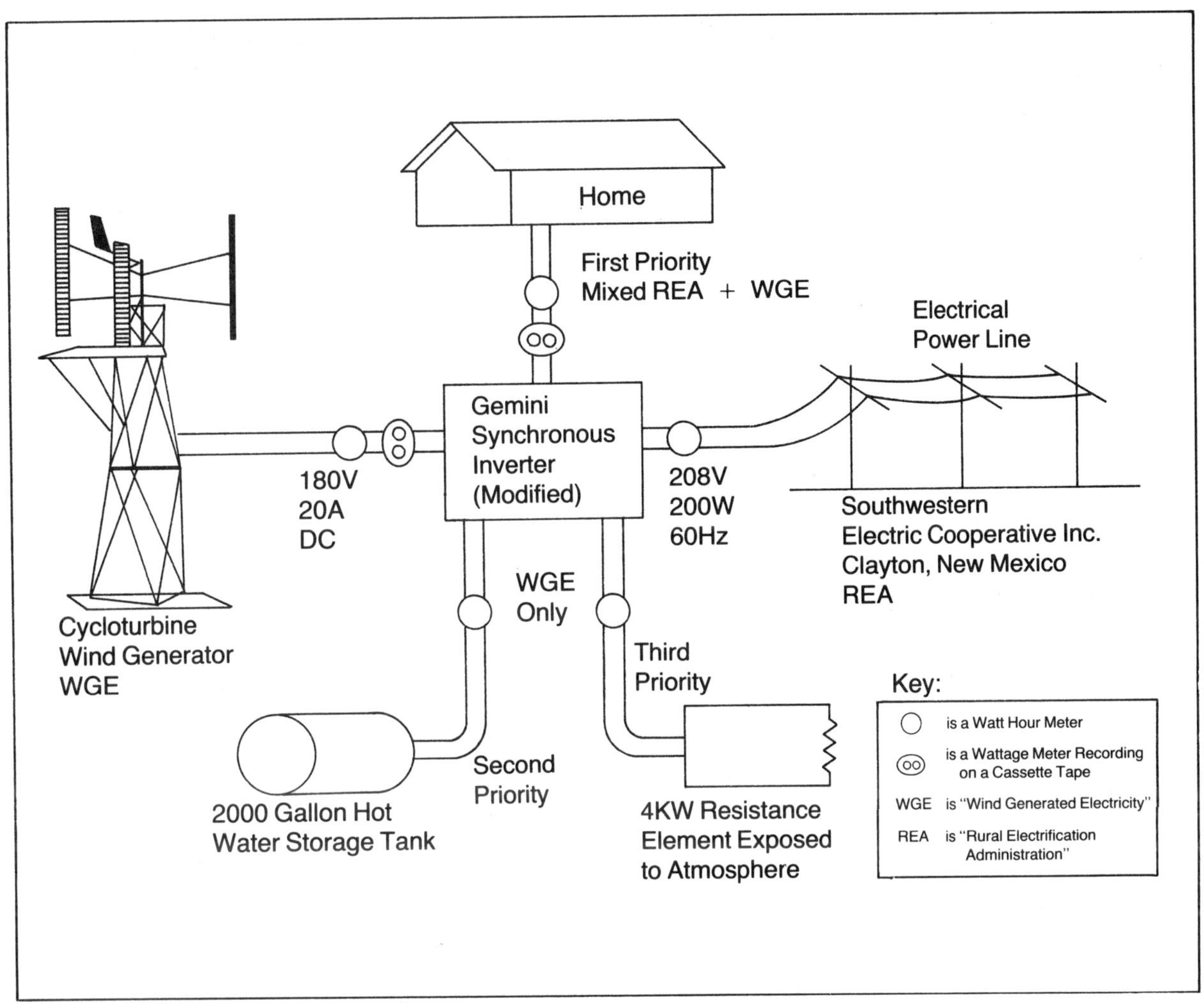

Fig. 4–4: Model electrical system/superintendent's home-livestock research station New Mexico State University. (Clayton, New Mexico)

remote mountaintops is costly. The electrical demand is very low due to the general use of solid-state, transistorized equipment. Much of the sophisticated engineering for these systems can be applied to the home owner's installation at lower costs.

NICKEL CADMIUM (Ni-Cad) BATTERIES

The chemistry and materials of Ni-Cad batteries are quite different from lead acid batteries and therefore perform in different ways. As the name implies, the plates are made of *nickel hydrate* and *cadmium oxide.* The electrolyte is a solution of potassium hydroxide rather than the sulfuric acid common to other types of batteries. The electrolytic chemical reaction causes the transfer of oxygen from one plate to another during the charging and discharging cycle. The electrolyte acts only as a medium for the transfer of hydroxl ions between positive and negative plates and does not vary with the state of the charge (high or low) in the battery. The important advantages of the Ni-Cad batteries are:

1. Plates are internally very strong and have an exceptionally long life.
2. The battery may stand idle for long periods of time in any state of charge or discharge and will suffer no damage from lack of use.
3. The battery is virtually free from self-discharge in temperate climates and therefore retains its charge for long periods of time.
4. Ni-Cads operate well under the widest of temperature ranges and the electrolyte will not freeze even at −50°F.

Most of the features of the Ni-Cad battery are especially desirable in wind energy systems. This is particularly true in very hot or cold climates. Battery usage in desert areas and in polar regions absolutely demands Ni-Cad battery banks for storage purposes. While Ni-Cads do lose some storage capacity under extreme cold, they will still function reasonably well, and allowances can be made for the reduced capacity. If we compare a lead acid battery with the Ni-Cad at −20°F (−29°C), assuming similar performance at 80°F (27°C), the Ni-Cad will retain about 65 percent of its original capacity while the lead acid battery will retain only about 30 percent of its original capacity. Further, the lead acid battery can freeze at −20°F if it is discharged at an 8-hour rate. However, if the discharge rate is reduced to the 20-hour rate, the Ni-Cad will retain about 80 percent of its capacity at −20°F.

In desert regions, where high temperatures of 120°F (49°C) are common, battery life is also affected. In the case of the lead acid battery the life expectancy is reduced to 20 percent of normal while the Ni-Cad will retain about 60 percent of its life expectancy. Demand for water by batteries increases with high temperatures. This is particularly true with the lead acid battery. Perhaps you are asking yourself, "If Ni-Cad's are so great, why aren't they recommended as standard equipment in a wind energy system?" There is one simple answer: money. They are very expensive.

Ni-Cads cost five to six times more than the equivalent lead acid battery. As inflation rages and lead prices keep going up, so does the nickel used in the Ni-Cad battery. Those who aren't dissuaded by the cost differential can maybe find some comfort in the thought that you always get what you pay for.

CONVERSION OF STORED DC POWER (USING DC POWER)

The stored energy in wind systems, as well as in other alternative schemes such as solar or hydro systems, is most often found in the form of DC, or direct current, from a battery source. The voltage from a DC battery source is an important consideration when it is desirable to use some of this energy in its existing form. For example, many readily available hand tools operate on either AC or DC. Known as universal appliances, these tools can use either kind of current without any conversion equipment. When the voltage is 120VDC, you can use these tools and thus avoid conversion problems.

There are some problems involved in using DC in the home. DC suffers substantial loss in the distribution system due to the resistance of the wiring coupled with the distance it must travel. This means that devices operating close to the DC source will have a measurably higher voltage than a device that is farther away. This distance does not have to be very great to reveal the effects of voltage loss. Another serious problem is that DC *cannot* be regulated for home use as AC can. Whatever the batteries are putting out in terms of voltage will drop, lights will get dimmer, motors will slow down, and current (amps) will increase. Wire sizes must be larger in diameter to accommodate the current fluctuations. This adds unnecessary cost to the system. In summary, the use of DC can be practical at or near the source but impractical as the distance from the power source increases.

One more important consideration is that DC moves only in one direction (it does not alternate as AC does) so that if a short circuit occurs, a great deal of heat is built up very rapidly causing wires to arc and burn. Electric arc welding is nothing more than putting the positive and negative terminals of a DC source close together without touching. The heat that develops is sufficient to melt steel. If you have ever watched someone handle an arc welder, you have a good idea of the intensity of the DC arc. You may also remember that if the welder *touched* the two terminals (the electrodes) together, they stuck and were very hard to pull apart. This happens because the current is always going in the same direction. In a home where DC is used for power, it is very important that all possible short circuits be eliminated. A short circuit can occur with AC, too, but since the current is changing (alternating) direction 60 times every second, a short can be disengaged more readily. The heating and arcing problem which happens when DC is used is the basic reason why AC circuit breakers and fuses should *not* be used in DC circuits unless the device specifically states that it may be used safely for both AC and DC. A DC circuit breaker is about twice the size (dimensional) of a comparable AC circuit breaker and is nearly twice the price. The reason for this is that as the breaker starts to disconnect the DC circuit, the current tries to continue to flow between the contact points and develops an arc like the arc of a welder. This arc must be stopped so that the contact points are not welded together or otherwise damaged. If this happens, the circuit breaker will no longer work. These features require space and money to accomplish, but they are essential. If an AC circuit breaker were to be used in a DC circuit, it is likely that the breaker contact points would be welded together the first time it is overloaded. The result would be a short circuit that raises the temperature along the distribution line until something burns up and breaks the connection. This "something" could be in the wall of the house, at the breaker box, or, worst of all, from an explosion caused by an internal spark igniting the hydrogen gas constantly being generated by the battery bank.

This information is not meant to scare or discourage you from using DC in your home. It is meant to make you fully aware of the problems that

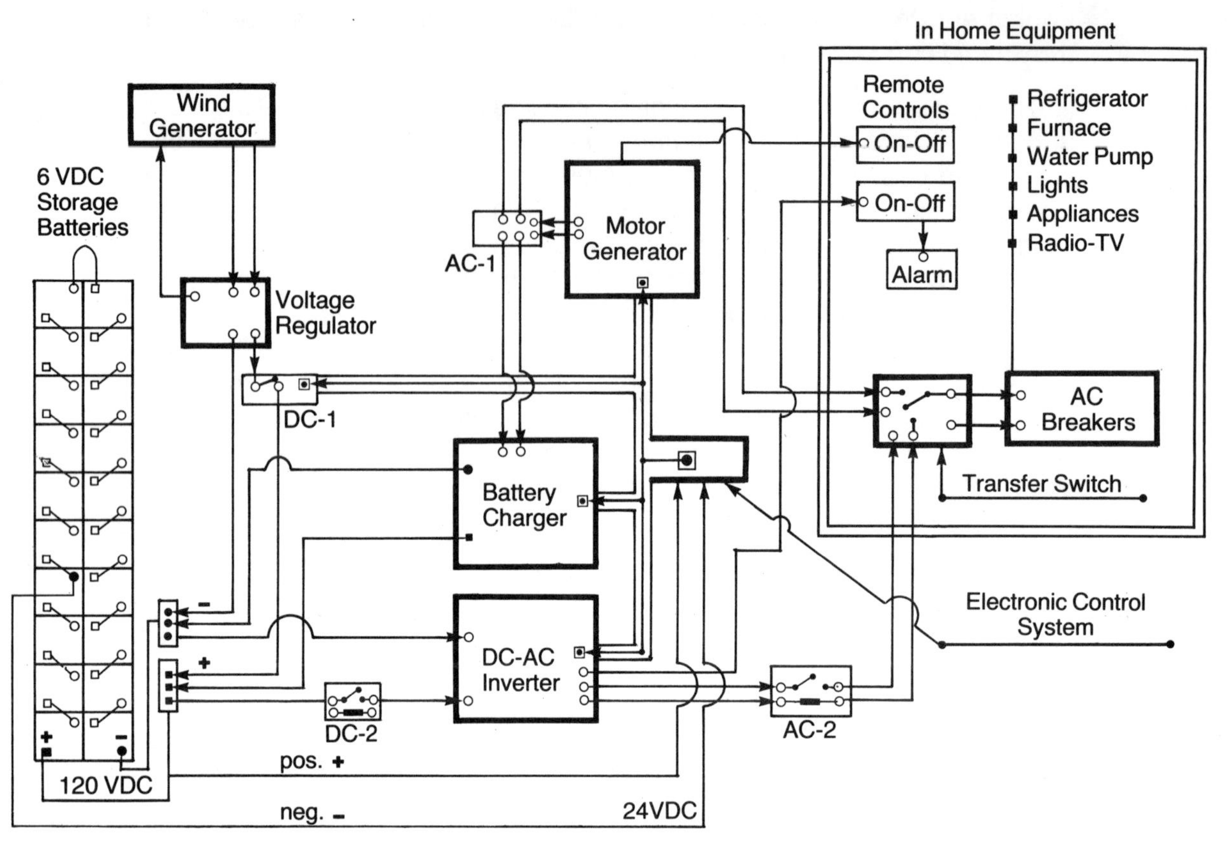

Fig. 4–5: A complete wind energy system.

sometimes arise with ill-conceived or poorly engineered systems. For many years we have been avoiding these problems because experts have known about them and have taken adequate preventative safety measures, such as building codes, to avoid them.

Fig. 4–5 diagrams a complete wind system, including a standby generator and battery charger with an electronic control device that does most of the system's management. This diagram presents an ideal system which may not be practical for all applications, but it is a starting point and does contain important items that need to be included in the most modest of wind energy systems. Although this is not a "how-to-do-it" book, it is important that you know what will be involved if you decide to become an active user of wind energy.

The system shown in the diagram will generally operate by itself. It will directly use energy from the wind-driven generator as long as the generator provides it. If the generator is creating power, and none of the power, or very little of it, is being used, the battery bank will be charged. If most of the stored energy from the batteries has been used and the batteries are nearly discharged, the *inverter* will start the motor-driven generator, which in turn will provide power for use in the total system or for battery recharging. When the batteries are fully charged, the charger will shut down the generator. If the generator does not start automatically, an alarm will sound, warning that the battery bank is down and that the generator has failed to start. Since the power for the alarm comes directly from the battery terminals, it will continue to sound for a long period of time (perhaps a month or more) or until the homeowner turns it off and corrects the problem with the generator. So even if you are away from home when this occurs, you will be made aware of the problem when you return. This kind of system eliminates the need for the homeowner to constantly monitor the condition of the battery to prevent overdischarge, which will damage the batteries. If the motor-driven generator and battery charger are eliminated from the system, the inverter can be designed to shut down when the batteries reach a state of complete discharge. This means that power to the home can be interrupted without warning and cannot be made operational until the wind-driven generator restores the batteries to a reasonable state of charge. Without a charger and a motor generator, management of the system and the direct observation of its condition must be done by the homeowner.

Another feature that can be added to this system turns off certain electrical loads in order of their priority when the battery reaches a certain low point. This conserves whatever energy is left in the battery bank until all stored energy has been used and all electrical loads have been disconnected.

A *load-shedding control system* has not been included in the system diagram because of the high cost of the controlling device. Even so, many systems have gone this far in their development because of the critical nature of the equipment used in the particular system. Essentially the device would be a microprocessor or small computer that takes into account factors of load, priority, and energy left in storage before it makes operating decisions. Built-in safety features include turn-off switches and fuses in the line to protect the devices in the system during operation. This also facilitates service of these devices in the event of failure or demand for maintenance. DC-1 in Fig. 4–5 protects the wind-driven generator from shorts or overloads and also cuts it off from the batteries while either the generator or the batteries are being serviced. DC-2 in the diagram (Fig. 4–5) protects the batteries from an overload or a short in the inverter or in the AC distribution lines to the home. Other devices are in the chain from the inverter to the house distribution. These units are safety features for additional protection. If all else fails, DC-2 (Fig. 4–5) will disconnect and protect the system. Both DC-1 and DC-2 are carrying DC and must be specifically designed and specified for DC use only as has been previously mentioned. Here there is no compromise.

AC-1 in Fig. 4–5 protects the generator from overload and shorts caused either by the charger or the house distribution AC breakers. The charger, and the inverter as well, nearly always have internal fuses to protect them from catastrophic failure. AC-2 in Fig. 4–5 protects the inverter from overloads and from shorts and also allows easy disconnection of the AC load for inspection for service or for battery maintenance. Both AC-1 and AC-2 carry AC and should be specifically designed and specified for AC application. Fuse sizes and types for these safety devices can be obtained from the equipment specifications data on the device itself.

The transfer switch located in the home can be used to feed electrical power to the home either from the batteries and the inverter or directly from the generator. This allows the entire wind system to be separated from the home during construction of the system or at those times when maintenance is required but electricity is still needed within the home. Remote controls for the generator and the inverter are provided in the home so that the owner may override the system and manually turn the units on or off.

DC TO AC INVERTERS

Since some devices in the home must operate on AC and the battery storage system is DC, it is necessary to convert the DC to AC in order to operate these devices. An inverter gets its name because it takes the continuous flow of DC and breaks it up into electric current that *alternates* first in one direction and then in the opposite direction. In the United States and in certain other countries in the western hemisphere, AC power is cycled at 60 times per second. In Europe it is cycled at 50 times per second. The choice of cycling (frequency of the AC) is a matter of arbitrary choice, and neither 50 nor the 60 cycles per second offers any particular advantage. Perhaps the United States chose the 60-cycle frequency because it is relative to the 60-minute hour and the 60-second minute, all of which are divisible by 10. The 50-cycle choice seems to parallel the metric system where everything is based on tens, hundreds, and thousands. Neither cycle presents special problems to the wind energy system owner.

However, AC power has some very distinct advantages over DC. It is quite easy to regulate or to change AC voltage, and it suffers less loss when it is distributed over an appreciable distance. Basic to this is the ability of AC to cut lines of flux in a transformer because of its change in direction. Alternating current in a coil of wire can, through a field of iron, transmit its power to another coil wound around the iron field in nearly direct proportion. The proportion of the second coil may be changed by increasing or decreasing its size, which raises or lowers the voltage. It is quite simple to take the 120 VAC output from an inverter or other AC source and step it up or down to a desired voltage. This is not possible with DC. Since many wind systems have the wind generator, the batteries, and the inverter some distance from where the electricity is being used, it would be desirable to step up the voltage at the generating source so that the distance to the home will not be a limiting factor in determining a suitable tower site.

The higher the AC voltage, the lower the current (amps). The lower the current, the lower the resistence (lower losses). The lower the current, the smaller the conducting wire must be; therefore, the lower the cost. So, if the distance between the house and the AC source is over 500 feet, it will usually prove advisable to step up the voltage at the source and step it down at the house for normal consumption. One might step up 120 VAC to 480 VAC for transmission and then return, or step down, to 120 VAC at the house. This amount of stepup will reduce the current the wire must carry to one-fourth of what it would be at 120 VAC. For example, if we are using 40 amps of power at 120 VAC at the house and the voltage between the house and the generating source is 480 VAC, then the current at 480 VAC would be 10 amps—a big difference. Certain types of transformers regulate the voltage as well as step it up or down. These transformers, however, are much more expensive and regulation should therefore be controlled at the AC source. This is usually accomplished by the inverter or the motor generator.

Making AC out of DC is not easy; nor is it inexpensive. The electronic parts in a static or electronic inverter must be able to withstand all of the power demands made in the home. It may seem strange, but power demands within the home are varied and sporadic. Careful consideration of these demands must be made so that proper equipment to do the job can be specified. Experience has shown that the motors used in the home are all too often underrated, inefficient, and generally very difficult to handle. The reason for this is that induction-type motors used in refrigerators, freezers, pumps, and furnaces require huge amounts of power to get started but need relatively little to keep running, This power drain can be as much as 10 times the amount needed to run the motor after it has been started. These starting *surges* of current put the electrical system and the inverter parts to the ultimate test. Even small refrigerators can raise current levels (amps) beyond the system's capability. This is particularly true when and if the batteries are 50 percent or more discharged at normal temperatures. This problem worsens when the temperature is near 0°F (18°C). In a wind system there is a limited supply of electrical power that goes "soft" when large electrical loads are present. The battery storage systems of wind energy sources will drop in voltage with high loads and become soft, unlike the power provided by a large utility company, which will continue to pour more energy into the line as the load increases. Keep in mind that as the battery voltage drops, the current (amps) goes up in an effort to supply the amount of power required.

For reasons of economy and system reliability it is wise to plan to use as few motors as possible and to be certain that the motors selected are efficient, are rated properly for the job, and are the *soft-start* or *capacitor-start* type.

ROTARY INVERTERS

DC may be converted to AC by means of a *rotary inverter* instead of a *static* or *electronic inverter.* The primary consideration is efficiency. Electronic inverters have efficiencies of up to 90 percent while rotary inverters at best are closer to 50 percent efficient. Rotary inverters are lower in cost and are generally more reliable, but this is not always true. In a limited power system, such as a wind generator system, the criterion is usually efficiency when choosing equipment. The rotary inverter is simply a DC-driven motor that generates AC power and is subject to frequency and voltage variations from 10 to 20 percent. Many of these inverters carry labels which clearly state that they are *not designed* to operate refrigerators or other induction-type motors but will operate *universal AC/DC motors* having brushes. Starting currents for rotary inverters can reach 400 amps for as long as 0.2 second.

SYNCHRONOUS INVERTERS

The name, *synchronous inverter,* implies that the inverter depends upon another AC source from which it derives its alternating frequency of 60 cycles per second. In wind energy system applications the inverters are connected to an existing power line. The presence of a utility power line also allows the owner to use the power company rather than batteries or one of the other schemes we have described earlier as the storage system. The synchronous inverter delivers power from the wind-driven generator or from the power company (Fig. 4–6). When the wind is generating power, that is the power to be used. When the wind is not blowing, electricity from the power company is used. If the wind is blowing and the system is not calling for power, the excess generated can be returned to the power company grid. In this kind of system the wind-generated power is supplemental, and the power company remains the prime source of energy. Since this system does not rely on batteries for storage, the cost of the system is considerably less than other energy systems. However, when the power company line fails, the home system is totally without power. The DC power is available, but only at the time it is being generated.

Because the synchronous inverter depends upon a connection with an existing AC power line, the installation must be approved by and coordinated with the local power company. The power company and the owner must set up some kind of fail-safe system to prevent the wind energy system from sending current down the line when the line is under repair and the repair man believes that all power is off. Since there are many of these systems in operation, we feel that this problem has been generally resolved, but because of the danger to individuals, this factor must continue to be considered as new systems are installed.

VOLTAGE REGULATION

Some inverters do not accurately regulate the AC voltage at its output point. These inverters should be avoided primarily because DC input from batteries varies widely. As the batteries discharge, the voltage drops. If the AC is not regulated at the inverter, then the AC voltage in the home will also drop as the voltage in the battery bank drops. In a 120 VDC battery system full-charge voltage is about 2.3 volts per cell. There are 60 2-volt cells at 2.3 VDC each, which equals 138 at full charge. When the batteries are fully discharged, the voltage is 1.75 volts per cell, which equals 105 VDC. This 33-volt difference can cause operating problems in AC devices if the voltage is not held at or near 120 volts. From our discussion about DC and AC, you will recall that AC is easily regulated while DC cannot be held close to any predetermined level. Regulation of about 5 percent is adequate for most home applications.

Fig. 4–6: Gemini Synchronous Inverter. (Windworks, Mukwanago, WI)

SPECIFICATIONS

SINGLE PHASE GEMINI				
Model	Power Capacity (kw)	Input Voltage (VDC)	Maximum Amperage (Amps)	Output Voltage (VAC)
PCU-40-1	2 4	0-100 0-200	20 20	120 240
PCU-80-1	4 8	0-100 0-200	40 40	120 240
PCU-150-1	15	0-200	75	240

THREE PHASE GEMINI				
Model	Power Capacity (kw)	Input Voltage (VDC)	Maximum Amperage (Amps)	Output Voltage (VAC)
PCU-200-3	20 20	250 500	80 40	240 480
PCU-400-3	40 40	250 500	160 80	240 480
PCU-500-3	50 50	250 500	200 100	240 480
PCU-1000-3	100 100	250 500	400 200	240 480

* U.S. patent Nos. 3,946,242 and 4,059,772

Specifications for inverters with capacities up to 1.5 megawatts available on request.

Source/Resource

BIBLIOGRAPHY

Books and Articles

Brangwyn, Frank and Preston, Hayter. *Windmills.* New York: Dodd, Mead, and Company, 1923.

Clark, Wilson. *Energy for Survival, the Alternative to Extinction.* Garden City, N.Y.: Doubleday, Anchor Press, 1974.

Clews, Henry. *Electric Power from the Wind.* East Holden, Maine: Solar Wind Company, 1973.

Dennis, Landt and Dennis, Lisl. *Catch the Wind.* New York: Four Winds Press, 1976. Price: $7.95. (Good general reference and pictorial history.)

Enertech Corporation. *Planning a Wind-Powered Generating System.* From the Enertech Corporation, Box 420, Norwick, Vermont 05055, 1977. (A manual written by one of America's leading companies concerning the design, supply, and installations of wind-powered generating systems. Both this manual and Henry Clews' *Electric Power from the Wind* may be ordered from Enertech for a total of $5.)

Freese, Stanley. *Windmills and Millwrighting.* Cambridge: At the University Press, 1957.

Golding, E. W. *The Generation of Electricity by Wind Power.* New York: Halstead Press, John Wiley & Sons, 1956. Price: $19.95. (A scholarly, theoretical treatment.)

Hackleman, Michael. *The Homebuilt, Wind-generated Electricity Handbook.* Maripose, California: Earthmind, 1975. Price: $8 postpaid. (Includes wind machine restoration, towers, installation, the control box, auxiliary electricity-generating equipment and wind machine design notes.)

———. *Wind and Windspinners.* Maripose, California: Earthmind, 1974. Price: $8 postpaid. (Emphasizes the savonius rotor by answering such questions as: How big a machine? Which generator? Where do I put it? What happens when the wind isn't blowing? How much electricity do I need? Do I have enough wind? Who can repair it? Can I get parts?)

Handbook of Homemade Power. The Mother Earth News, c/o The Register and Tribune Syndicate, Inc., 715 Locust St., Des Moines, Iowa 50304.

Hopkins, R. T. and Freese, S. *In Search of English Windmills.* London: Cecil Palmer Publishing Company, 1931.

Leckie, Jim, et. al. *Other Homes and Garbage.* San Francisco: Sierra Club, 1975. Price: $9.95. (ISBN: 87156–141–7)

Livingstone, Richard N., "The Search for Alternative Energy." *The New Englander* 22, no. 8 (December 1975): 24–30.

Meyer, Hans. "Wind Generators: Here's an Advanced Design You Can Build." *Popular Science* 201, no. 5 (November 1972): 103–105.

Park, Jack. *Simplified Wind Power Systems for Experimenters,* 2nd ed. Sylmar, California: Helion, Inc., 1975. Price $6 postpaid. (Written by an engineer; information on windmill design methods for non-engineers.)

Putnam, Palmer Cosslet. *Energy in the Future.* New York: D. Van Nostrand Co., 1953.

———. *Power from the Wind.* New York: D. Van Nostrand, 1948.

Reed, Jack. "Wind Power Climatology of the United States." *Weatherwise,* December 1976. *Weatherwise* is a publication of the American Meteorological Society.

Reynolds, John. *Windmill and Watermills.* New York: Praeger Publishers, 1975.

Solar Energy Applications Laboratory. *Energy from the Wind—An Annotated Bibiliography,* Fort Collins, Colorado: Colorado State University, 1975, supplement 1977. Price: $15 postpaid.

Spier, Peter. *Of Dikes and Windmills.* New York: Doubleday, 1969.

Stokhuyzen, Frederick. *The Dutch Windmill.* New York: Praeger, 1970.

Stoner, Carol Huping, ed. *Producing Your Own Power.* New York: Vintage Books, 1975. Price: $3.95.

Torrey, Volta. *Wind Catchers—American Windmills of Yesterday and Tomorrow.* Brattleboro, Vermont: Stephen Greene Press, 1976. Price: $12.95. (A broad, historical summary with interesting anecdotes.)

Vince, John. *Discovering Windmills.* Aylesbury, England: Shire Publications, 1973.

Wailes, Rex. *English Windmills.* London: Routledge and Kegan Paul, 1954.

———. *Windmills in England.* London: Architectural Press, 1948.

Wind Energy Utilization. Albuquerque: Technology Application Center, University of New Mexico.

Wind Machines (NSF–RA–N–75–051). Washington: U.S. Government Printing Office, 1976. Price: $2.25. (A comprehensive technical, historical, and economic summary.)

Reports

AAI Corporation and Institute of Gas Technology. *Production of Methane Using Offshore Wind Energy.* R. B. Young, A. F. Tiedeman, Jr., T. G. Marianawski, E. H. Camara, November 1975. Contract No. NSF-C993. Final Report: (PB 252 307), 131 pp. Executive Summary: (PB 252 308), 29 pp.

Alaska, University of, Geophysical Institute. *Study of Alaskan Wind Power and Its Possible Applications* (Final Report, May 1, 1974–January 30, 1976). T. Wentink, Jr., February 1976, 139 pp. Contract No. NSF–AER–74–00239 (NSF/RANN/SE/AER–74–00239) (PB 253 339).

———. *Wind Power Potential of Alaska. Part II, Wind Duration Curve Fits and Output Power Estimates for Typical Windmills.* T. Wentink, Jr., August 1976, 86 pp. Contract No. E(45–1)–2229. RLO/2229/T12–16/1).

Battelle Memorial Institute, Columbus Laboratories. *An Evaluation of the Potential Environmental Effects of Wind Energy Systems Development* (Final Report). S. Rogers, et. al., August 1976. Contract No. NSF–AER–75–07378. (ERDA/NSF/07378–75/1).

Battelle-Pacific Northwest Laboratories. *Annual Report of the Wind Characteristics Program Element for the Period April 1967–June 1977.* J. V. Ramsdell, June 1977. Contract No. EY–76–C–06–1830. (BNWL–2220–WIND–10).

Colorado State University. *Sites for Wind Power Installations: Wind Tunnel Simulation of the Influence of Two-dimensional Ridges on Wind Speed and Turbulence* (Annual Report). R. N. Meroney, et. al., July 1976, 80 pp. Contract No. NSF–RANN–GAER–75–00702. (ERDA/NSF/00702–75/1).

———. *Wind-Powered Aeration for Remote Locations* (Final Report, March 15, 1975–August 31, 1976). P. M. Schierholz, October 1976, 130 pp. Contract No. NSF–G–AER–75–00833. (ERDA/NSF/00833–75/1).

Dayton, University of, Research Institute. *Electrofluid Dynamic (EFD) Wind Driven Generator.* J. E. Minardi, M. O. Lawson, G. Williams, October 1976. Contract No. EY–76–S–02–4130.

Energy Research and Development Administration. *Federal Wind Energy Program, Summary Report,* January 1, 1977, 56 pp. U.S. Government Printing Office (Stock No. 060–000–00048–4).

———. *Federal Wind Energy Program, Summary Report.* Division of Solar Energy, October 1975, 78 pp. (ERDA–84).

General Electric, Space Division. *Design Study of Wind Turbines 50 kW to 3000 kW for Electric Utility Applications.* December 1976. Volume I (Summary Report): NASA CR–134934; Volume II: NASA CR–134035; Volume III: NASA CR–134936.

———. *Wind Energy Mission Analysis,* February 1977. Contract No. E(11–1)–2578. Executive Summary: C00/2578–1/1, 2 pp.; Final Report: C00/2578–1/2, 219 pp. Appendices A–J: COO/2578–1/3, 480 pp.

George Washington University. *Legal-Institutional Implications of Wind Energy Conversion Systems, Final Report.* L. H. Mayo, et. al., September 1977, 333 pp. Contract No. APR 75–19137 (NSF/RA–770204).

Georgia Institute of Technology. *Reference Wind Speed Distributions and Height Profiles for Wind Turbine Design and Performance Evaluation Applications.* C. G. Justus, W. K. Hargraves, A. Mikhail, August 1976, 96 pp. Contract No. E(40–1)–5108 (ORO/5108–76/4).

———. *Wind Energy Statistics for Large Arrays of Wind Turbines (New England and Central U.S. Regions).* C. G. Justus, August 1976, 129 pp. Contract No. NSF–AER75–00547 (PB 260 679).

Grumman Aerospace Corporation. *Investigation of Diffuser-Augmented Wind Turbines.* R. A. Oman, January 1977. Contract No. EY–76–C–02–2616. Executive Summary: (COO/2616–1). Technical Report: (COO/2616–2).

Institute of Gas Technology. *Wind-Powered Hydrogen-Electric Systems for Farm and Rural Use.* J. B. Pangborn, April 1976, 158 pp. Contract No. NSF–AER–75–00772 (PB 259 318).

JBF Scientific Corporation. *Summary of Current Cost Estimates of Large Wind Energy Systems* (Special Technical Report), February 1977, 62 pp. Contract No. E(49–18)–2364 (DSE/2521–1).

Lockheed-California Company. *100 kW Metal Wind Turbine Blade Basic Data, Loads, and Stress Analysis.* A. W. Cherritt and J. A. Gaidelis, June, 1975. NASA Contract No. NAS3–19235 (NASA CR–134956).

———. *100kW Metal Wind Turbine Blade Dynamics Analysis, Weight/Balance and Structural Test Results.* W. D. Anderson, June 1975. NASA Contract No. NAS3–19235. (NASA CR–134957).

———. *Wind Energy Mission Analysis,* October 1976; Contract No. EY–76–C–03–1075. Executive Summary: SAN/1075–1/3, 30 pp.; Final Report: SAN/1075–1/1; Appendix: SAN/1075–1/2.

Martin Marrietta Laboratories. *Segmented and Self-Adjusting Wind Turbine Rotors* (Final Report). P. F. Jordon, R. L. Goldman, April 1976, 113 pp. Contract No. EY–76–C–02–2613 (COO/2613–2).

Massachusetts Institute of Technology. *Research on Wind Energy Conversion Systems.* R. H. Miller, December 1976. Contract No. NSF–AER–75–00826.

Massachusetts, University of, Amherst. *Investigation of the Feasibility of Using Windpower for Space Heating in Colder Climates* (Third quarterly progress report covering the final design and manufacturing phase of the project. September–December 1975). W. E. Heronemus, December 1975, 165 pp. Contract No. NSF–AER–75–00603 (ERDA/NSF/00603–75/T1).

McDonnell Aircraft Co. *Feasibility Investigation of the Gyromill for Generation of Electrical Power.* R. V. Brulle, November 1975, 155 pp. Contract No. E(11–1)–2617 (COO–2617–75/1).

Michigan State University, Division of Engineering Research, *Application Study of Wind Power Technology to the City of Hart, Michigan.* J. Asmussen, P. D. Fisher, G. L. Park, O. Krauss, December 1975, 103 pp. Contract No. E(11–1)–2603 (COO–2603–1).

Michigan, University of. Radiation Laboratory *TV and FM Interference by Windmills* (Final Report). T.B.A. Senior, et. al., February 1977, 150 pp. Contract No. EY–76–S–02–2846 (COO/2846–76/1).

Mitre Corporation. *Wind Energy Conversion Systems, Proceeding of the Second Workshop* (on June 9–11, 1975). F. R. Eldridge, June 1975, 536 pp. Contract No. NSF–AER–75–12937 (NSF–RA–N–75–050).

———. *Wind Machines.* F. R. Eldridge, October 1975, 84 pp. (NSF–RA–N–75–051) U.S. Government Printing Office (Stock No. 038–000–00272–4).

NASA-Lewis Research Center. *Benefit-Cost Methodology Study with Example Application of the Use of Wind Generators.* R. P. Zimmer, C. G. Justus, R. N. Mason, S. L. Robinette, P. G. Sassone, W. A. Schaffer of Georgia Institute of Technology, July 1975, 411 pp. (NASA CR–134864).

———. *Installation and Initial Operation of a 4100 Watt Wind Turbine.* H. B. Tyron, T. Richards, December 1975. (NASA TM X–71831).

———. Interagency Agreement No. E(49–26)–1028. *Free Vibrations of the ERDA-NASA 100 kW Wind Turbine.* C. C. Chamis, T. L. Sullivan, February 1976. (NASA TM X–71879).

———. *Transient Analysis of Unbalanced Short Circuits of the ERDA–NASA 100 kW Wind Turbine Alternator.* H. H. Hwang, Leonard J. Gilbert, July 1976. (NASA TM X–73459).

———. *Early Operation Experience on the ERDA/NASA 100 kW Wind Turbine.* J. C. Glasgow, B. S. Linscott, September 1976 (NASA TM X–71601).

———. *Tower and Rotor Blade Vibration Test Results for a 100 kW Wind Turbine.* B. S. Linscott, W. R. Shapton, D. Brown, October 1976 (NASA TM X–3426).

———. *Wind Tunnel Measurements of the Tower Shadow on Models of the ERDA/NASA 100 kW Wind Turbine Tower.* J. M. Savino and L. H. Wagner, November 1976 (NASA TM X–73548).

———. *Synchronization of the ERDA–NASA 100 kW Wind Turbine Generator with Large Utility Networks.* H. H. Hwang and L. J. Gilbert, March 1977, 17 pp. (NASA TM X–73613).

———. *Vibration Characteristics of a Large Wind Turbine Tower on Non-Rigid Foundations.* S. T. Yee, T. Yung, P. Change, et. al. May 1977 (ERDA/NASA 1004–77/1).

———. *Preliminary Design of a 100 kW Turbine Generator.* R. L. Puthoff, P. J. Sirocky, 1974, 22 pp. (NASA TM X–71585; E–8037) (N–74–31527).

———. *Structural Analysis of Wind Turbine Rotor for NSE-NASA MOD-O Wind-Power System.* D. A. Spera, March 1975, 39 pp.(NASA TM X–3198; E–8133) (N–75–17712).

———. *Plans and Status of the NASA-Lewis Research Center Wind Energy Project.* R. Thomas, R. Puthoff, J. Savino, W. Johnson, 1975, 31 pp. (NASA TM X–71701; E–8309) (N–75–21795).

———. *A 100 kW Experimental Wind Turbine: Simulation of Starting Overspeed and Startdown Characteristics.* L. Gilbert, February 1976 (NASA TM X–71864).

———. *Large Experimental Wind Turbines—Where We Are Now.* R. L. Thomas, March 1976 (NASA TM X–71890).

———. *Fabrication and Assembly of the ERDA/NASA 100 kW Experimental Wind Turbine.* R. L. Puthoff, April 1976 (NASA TM X–3390).

———. *Design Study of Wind Turbines 50 kW to 3000 kW for Electric Utility Applications* (See Technology Development).

———. *Wind Energy Conversion Systems, Workshop Proceedings* (Washington, D.C. June 11–3). J. M. Savino, December 1973, 258 pp. Grant No. NSF–AG465 (NSF–RA–N–73–006) (PB 231 341).

———. *Dynamic Blade Loading in the ERDA/NASA 100 kW and 200 kW Wind Turbines.* D. A. Spera, D. C. Janetzke, T. R. Richards, May 1977 (ERDA/NASA 1004–77/2).

———. *Drive Train Normal Modes Analysis for the ERDA/NASA 100-Kilowatt Wind Turbine Generator.* T. L. Sullivan, D. R. Miller, D. A. Spera, July 1977 (ERDA/NASA/1028–77–1).

———. *Investigation of Excitation Control for Wind Turbine Generator Stability.* V. D. Gebben, August 1977 (ERDA/NASA 1028–77/3).

———. *Nastran Use for Cyclic Response and Fatigue Analysis of Wind Turbine Towers.* C. C. Chamis, P. Manos, J. H. Sinclair, J. R. Winemiller, October 1977, 20 pp. (ERDA/NASA 1004–77/3).

———. *Wind Energy Utilization, A Bibliography.* Technical Applications Center, University of New Mexico, for NASA–LeRC. (TACW–75–700).

NOAA-National Climatic Center. *Initial Wind Energy Data Assessment Study.* M. J. Changery, May 1975, 132 pp. Contract No. NSF–AG–517 (NSF–RA–N–75–020) (PB 244 132).

Oklahoma State University. *Development of an Electrical Generator and Electrolysis Cell for a Wind Energy Conversion System* (Final Report, July 1, 1973–July 1, 1975). W. Hughes, H. J. Allison, R. G. Ramarkumar, July 1975, 280 pp. Contract No. NSF–AER–75–00647 (NSF/RA/N–75–043) (PB 243 909).

Oregon State University. *Aerodynamic Performance of Wind Turbines.* R. E. Wilson, P. B. S. Lissaman, S. N. Walker, June 1976, 170 pp. Contract No. NSF–AER–74–04014 A03 (PB 259 089).

———. *Applied Aerodynamics of Wind Power Machines.* R. E. Wilson, P. B. S. Lissaman, July 1974, 116 pp. Contract No. NSF–AER–74–04014 A03 (PB 238 595).

Paragon Pacific, Inc. *Coupled Dynamics Analysis of Wind Energy Systems,* February 1977. NASA Contract No. NAS3–197707 (NASA CR–135152).

Princeton University. *Optimization and Characteristics of a Sailwing Windmill Rotor.* M. D. Maughmer, March 1976. Contract No. GI–41891 (NSF/RANN/GI–41891/FR/75/4).

Sandia Laboratories, Contract No. AT(29–1)–789. *Vertical-Axis Wind Turbine Technology Workshop* (held at Sandia Laboratories, Albuquerque, New Mexico, May 18–20, 1976). L. Wetherholt, July 1976, 439 pp. (SAND76-5586).

———. *The Vertical-Axis Wind Turbine—How it Works.* B. F. Blackwell, April 1974, 8 pp. (SLA–74–0160).

———. *Blade Shape for a Troposkien Type of Vertical-Axis Wind Turbine.* B. F. Blackwell, G. E. Reis, April 1974, 24 pp. (SLA–74–0154).

———. *An Electrical System for Extracting Maximum Power from the Wind.* A. F. Veneruso, December 1974, 29 pp. (SAND74–0105).

———. *Some Geometrical Aspects of Troposkiens as Applied to Vertical-Axis Wind Turbines.* B. F. Blackwell, G. E. Reis, March 1975. (SAND74–0177).

———. *Practical Approximations to a Troposkien by Straight-Line and Circular-Arc Segments.* G. E. Reis, B. F. Blackwell, March 1975, 34 pp. (SAND74–0100).

———. *Wind Energy—A Revitalized Pursuit.* B. F. Blackwell, L. V. Feltz, March 1975, 16 pp. (SAND75–0166).

———. *An Investigation of Rotation-Induced Stresses of Straight and of Curved Vertical-Axis Wind Turbine Blades.* L. V. Feltz, B. F. Blackwell, March 1975, 20 pp. (SAND74–0379).

———. *Application of the Darrieus Vertical-Axis Wind Turbine to Synchronous Electrical Power Generation.* J. F. Banas, E. G. Kadlec, W. N. Sullivan, March 1975, 14 pp. (SAND75–0165).

———. *Nonlinear Stress Analysis of Vertical-Axis Wind Turbine Blades.* W. I. Weingarten, R. E. Nickell, April 1975, 20 pp. (SAND74–0378).

———. *Methods of Performance Evaluation of Synchronous Power Systems Utilizing The Darrieus Vertical-Axis Wind Turbine.* J. F. Banas, E. G. Kadlec, W. N. Sullivan, April 1975, 22 pp. (SAND75–0204).

———. *The Darrieus Turbine: A Performance Prediction Model Using Multiple Streamtubes.* J. H. Strickland, October 1975 (SAND75–0431).

———. *Engineering of Wind Energy Systems.* J. F. Banas, W. N. Sullivan, January 1976 (SAND75–0530).

———. *Wind Tunnel Performance Data for the Darrieus Wind Turbine with NACA–0012 Blades.* B. F. Blackwell, L. V. Feltz, R. E. Sheldahl, 1976 (SAND76–0130).

———. *Wind Tunnel Performance Data for Two- and Three-Cup Savonius Rotors.* B. F. Blackwell, L. V. Feltz, R. E. Sheldahl, July 1977, 108 pp. (SAND76–0131).

———. *Engineering Development Status of the Darrieus Wind Turbine.* B. F. Blackwell, W. N. Sullivan, R. C. Reuter, J. F. Banas, March 1977, 68 pp. (SAND76–0650).

———. *Darrieus Vertical-Axis Wind Turbine Program at Sandia Laboratories.* E. G. Kadlec, August 1976, 11 pp. (SAND76–5712).

———. *Status of the ERDA/Sandia 17-Metre Darrieus Turbine Design.* B. F. Blackwell, September 1976, 16 pp. (SAND76–5683).

Sandia Laboratories, Contract No. S189–76–32. *Wind Energy Potential in New Mexico.* J. W. Reed, R. C. Maydew, B. F. Blackwell, July 1974, 40 pp. (SAND74–0071).

———. *Wind Power Climatology.* J. W. Reed, December 1974 (SAND74–0435).

———. *Wind Power Climatology of the United States.* J. W. Reed, May 1975, 163 pp. (SAND74–3078).

Societal Analytics Institute, Inc. *Barriers to the Use of Wind Energy Machines: The Present Legal/Regulatory Regime and a Preliminary Assessment of Some Legal/Political/Societal Problems.* R. F. and H. J. Taubenfeld, July 1976, 159 pp. Contract No. NSF–AER75–18362 (PB–263 567).

United Technologies Research Center. *Self-Regulating Composite Bearingless Wind Turbine* (Final Report, June 3, 1975–June 2, 1976). M. C. Cheney; P. A. M. Spierings, September 1976, 62 pp. Contract No. EY–76–C–02–2614 (COO/2614–76/1). Executive Summary, 13 pp. (COO/2614–76/2).

West Virginia University. *Innovative Wind Machines* (Executive Summary and Final Report). R. E. Walters, et. al., June 1976. Contract No. EY–76–C–05–5135 (ERDA/NSF/00367–76/2).

Other Sources of Information

Alternative Sources of Energy
Rt. 2, Box 90A
Milaca, MN 56353
(Quarterly)

Alternative Sources of Energy
Rt. 1, Box 36B
Minong, WI 54859

De Hollandsche Molen
9 Reguliersgracht
Amsterdam-C
Netherlands

Earthmind
5249 Boyer Road
Mariposa, CA 95338

Electrical Research Association
Cleeve Road
Leatherhead, Surrey
England

Garden Way Laboratories
Charlotte, VT 05445

Helion, Inc.
Box 445
Brownsville, CA 95919

McGill University
Macdonald College
Brace Research Institute
Ste. Anne de Bellevue 800
Quebec, Canada

Mother Earth News
Box 70
Hendersonville, NC 28739

NASA Lewis Research Center
Technology Utilization Office
21000 Brookpark Road
Cleveland, OH 44135
(216) 433–4000, ext. 6833/6832
Large, horizontal-axis wind generators.

National Bureau of Standards
Office of Energy-Related Inventions
George Lewitt, Chief
Washington, DC 20234
Evaluation of wind energy-related ideas or inventions.

National Climatic Center
Federal Office Building
Asheville, NC 28801
Local wind speed records.

National Technical Information Service
Springfield, VA 22161
(703) 321–8500
Reports and bibliographies of federal government energy-related activities.

New Mexico State University
Las Cruces, NM 88001

Newsletter
Wind Energy Society of America
1700 East Walnut
Pasadena, CA 91106
(Quarterly. Price: $20 per year.)

Rockwell International
Rocky Flatts Plant
Lou Seaverson
P.O. Box 464
Golden, CO 80401
(303) 497–4943/497–4470

Sandia Laboratories
Division 5628
Ben Blackwell
P.O. Box 5800
Albuquerque, NM 87115
Vertical-axis (Darrieus) wind turbines

Solar Wind
Henry Clews
RFD 2, Happytown Rd.
East Holden, ME 04429

Sunflower Power Co.
Rt. 1, Box 93A
Oskaloosa, KS 66066

United States Department of Commerce
National Technical Information Service
Energy Research and Energy Information Services
Washington, DC 20004

University of New Mexico
Technology Application Center
Albuquerque, NM 87131
(505) 277–3622
Reports and bibliographies of NASA and other energy-related research.

VITA (Volunteers in Technical Assistance)
Publication Services
3706 Rhode Island Ave.
Mt. Rainier, MD 20822

The Wind Power Digest
American Wind Energy Association
Bristol, IN 46507
(Quarterly. Price: $6 per year.)

Windustries
Great Plains Windustries
Box 126, Lawrence, KS 66044
(Quarterly. Price: $10 per year.)

Windworks
Rt. 3, Box 329
Mukwonago, WI 53149

United States Department of Energy
Wind Energy Program
Louis V. Divone, Manager
Mail Station 3–168
20 Massachusetts Avenue, N.W.
Washington, DC 20545
(202) 376–4460
Federal government wind energy research programs.

United States Department of Energy
Division of Distributed Solar Technology
Office of the Director
Washington, DC 20545

University of Massachusetts
Department of Mechanical Engineering
Amherst, MA 01002

Source/Resource

SUPPLIERS

Aeolian Kinetic™
P.O. Box 100
Providence, RI 02901
(401) 274-3690

Aerospace Systems, Inc.
1 Vinebrook Park
Burlington, MA 01803

Advance Industries
2301 Bridgeport Drive
Sioux City, Iowa 51102

Aeroelectric
13517 Winter Lane
Cresaptwon, MD 21502

Aeropower
2398 4th Street
Berkeley, CA 94710

Aerowatt
37, Rue Chanzy 75–Paris 11°
France

Aircraft Components
North Shore Drive
Benton Harbor, MI 49022

Alternative Energy Services Ltd.
2 Circular Road
Douglas, Isle of Man
England

Aluminum Company of America Labs
Alcoa Laboratory
Alcoa Center, PA 15069

Amerenalt Corporation
Box 905
Boulder, CO 80302

American Tower Company
Shelby, OH 44875
(419) 347-1185

American Wind Turbine Co., Inc.
1016 East Airport Road
Stillwater, OK 74074

P. Andrag & Sons Ltd.
P.O. Box 364
Belleville, Cape Province
South Africa

Astro Research Corporation
P.O. Box 4128
Santa Barbara, CA 93103
(805) 684-6640

ATR Electronics, Inc.
300 East 4th Street
St. Paul, MN 55101
(612) 222-3791

Automatic Power, Inc.
P.O. Box 18738
Houston, TX 77023

A.W.E.
Box 142
South Chelmsford, MA 01824

Bendix Environmental Science Division
1400 Taylor Avenue
Baltimore, MD 21204

Best Energy Systems for Tomorrow, Inc.
Route 1, Box 106
Necedah, WI 54646
(608) 565-7200

Booger County Enterprises
Star Route
Witter, AR 72766

Robert Bosch
33 Osborne Road
Southsea
Portsmouth, Hampshire
England

Bucknell Engineering Company, Inc.
10717 East Rush Street
South El Monte, CA 91733

Budgen & Associates
72 Broadview Avenue
Pointe Claire 710
Quebec, Canada

C & D Batteries
3404 Walton Road
Plymouth Meeting, PA 19462
(215) 828-9000

Carter Motor Company
2711 West George St.
Chicago, IL 60618
(312) 588–7700

C. F. Casella & Co.
Regent House
Britannia Walk, London N.1.
England

Century Storage Battery Company, LTC
Birmingham Street
Alexandria, Australia

Chalk Wind Systems
P.O. Box 446
St. Cloud, FL 32769

Chloride Transipack Ltd.
Stanley Road
Bromley, Kent
England

Clean Energy Products
3534 Bagley, N.
Seattle, WA 98103

Climet, Inc.
1620 West Colton Ave.
Redlands, CA 92373
(714) 793–2788

Cojo Wind Company
Hollister Ranch No. 6
Gaviota, CA 93017

P. P. Controls Ltd.
Crosslands Road
Hounslow, Middlesex
England

Coulson Wind Electric
RFD 1, Box 225
Polk City, IA 50226

Crowdis Conservers
RR 3, MacMillan Mt.
Cape Breton, Nova Scotia
Canada BOE 1B0

Dean Bennet Supply Company
4725 Lipan Street
Denver, CO 80211

Delatron Systems Corporation
553 Lively Boulevard
Elk Grove Village, IL 60007
(312) 438–9225 or 593–2270

Dempster Industries
P.O. Box 848
Beatrice, NB 68310

Domenico Sperandio & Ager
Via Cimarosa 13–21
58022 Folloncia (GR)
Italy

Dominion Aluminum Fabricators
3570 Hawkstone Rd.
Mississauqa, Ontario L5C 2V8
Canada

Dunlite Electrical Products Co.
Division of Pys Ind.
28 Orsmond St.
Hindmarch, S.
Australia

Dwyer Instruments, Inc.
P.O. Box 373
Michigan City, IN 46360
(219) 872–9141

Dynamote Corporation
1130 N.W. 85th Street
Seattle, WA 98117
(206)784–1900

Earth Mind
2651 O'Josel Drive
Saugus, CA 91350

Edmond Scientific Company
380 EDS Corp. Bldg.
101 East Gloucester Pike
Barrington, NY 08007

Eldon Arms
Box 7
Woodman, WI 53827

Elektro G.m.b.H.
Winterthur (Schweiz)
St. Gallerstrasse 27
8400 Winterthur
Switzerland

Elgar Corporation
8225 Mercury Court
San Diego, CA 92111
(714) 565–1155

Empire Energy Development Corp.
3371 West Hampden Ave.
Englewood, CA 80110

ENAG S.A.
Rue de Pont l'Abbe
Quimper (Finistere)
France

Energy Alternatives
Box 223
Leverett, MA 01054

52 French King Highway
Greenfield, MA 01301

Energy Development Company
179E Road No. 2
Hamburg, PA 19526
(215) 562–8856

Energy—2000
Route 800, RFD No. 3
Winstead, CT 06098

Enertech Corporation
P.O. Box 420
Norwich, VT 05055

Environmental Design & Construction
RFD
E. Wallingford, VT 05742

Environmental Energies, Inc.
P.O. Box 73
Front Street
Copemish, MI 49625

21243 Grand River
Detroit, MI 48219

Environmental Resource Group
Box 3A, RD 2
Williston, VT 05495

Fenton's Feeders
Route 1, Box 124
Arcadia, FL 33821
(817) 494–2727

Flanagan's Plans
2032 23rd St.
Astoria, NY 11105

P.O. Box 891
Cathedral Station
New York, NY 10025

Forrestal Campus Library
Princeton University
Princeton, NJ 08540

4-Winds of Alaska
5100 Vi Street
Anchorage, AK 99507

Garden Way Laboratories
Charlotte, VT 05445

Gemini Sychronous Inverter
Windworks
Box 329, Route 3
Mukwanago, WI 53149
(414) 363–4088

Georator Corporation
Box 70
9016 Prince William Street
Manassas, VA 22110
(703) 368–2101

Globe Battery
Division of Globe Union, Inc.
5757 North Green Bay Avenue
Milwaukee, WI 53201
(414) 228–2581

Gould Inc.
Industrial Battery Division
2050 Cabot Boulevard West
Langhorne, PA 19047
(215) 752–0555

Grumman Energy Systems
4175 Veterans Memorial Highway
Ronkonkoma, NY 11779

Gurnard Mfg. Corp.
100 Airport Road
Beverly, MA 01915

Helical Power Ltd.
36 Fontwell Drive
Glen Parva, Leicester LE2 9ML
England

Heller-Aller Co.
Corner Perry & Oakwood
Napoleon, OH 43545

Homecraft
2350 West 47th St.
Denver, CO 80211

Ralph Howe Marketing Ltd.
New Orchard and High Street
Poole, Dorset
England

Independent Energy Company
314 Howard Avenue
Ewarthmore, PA 19081

Independent Power Developers
Box 618
Noxon, MT 59853

Industrial Instruments Ltd.
Stanley Rd.
Bromley BR2 9JF
Kent, England

Jack Park
Box 4301
Sylmar, CA 91342

Johnson Division of UOP Co., Inc.
P.O. Box 3118
St. Paul, MN 55165

Kahl Scientific Instrument Corporation
P.O. Box 1166
El Cajon (San Diego), CA 92022
(714) 444–2158

Kedco, Inc.
9016 Aviation Blvd.
Inglewood, CA 90301

Kemah Power Company
Box 776–633 Pleasant Ave.
Saugatuck, MI 49453

Kramco
P.O. Box 1536
Allentown, PA 18105
(215) 437–6758

Lane & Bowler Company
P.O. Box 15528
Houston, TX 77020
(713) 672–7561

Living Energy Consultants
131 Beaver Ave.
Colorado Springs, CO 80906

Low Energy Systems
3 Larkfield Gardens
Dublin 6
Ireland

Low Impact Technology
73 Molesworth Street
Wadebridge
Cornwall, England

Lubing
Maschiningsbrik, Ludwig Bening
P.O. Box 171
D–2847 Barstorf, Vasttyskland
Germany

Lubing Maschinenfabrik
2847 Barnstort
Postfach 110
West Germany

Joseph Lucas Ltd.
Great Hampton Street
Birmingham
England

McDonnel Aircraft
Box 516
St. Louis, MO 63166

M & B Alternate Energy Systems
Dell Acres, Part A
Pierre, SD 57501

Makia Ocean Engineering
Box 1194
Kailua, Oahu, Hawaii 96734
(808) 259–5904 or 259–5722

Maschinenfabrik Ludwig Bening
Postfach 171
2847 Barnsdorf
West Germany

Massawippi Wind Electric Co.
R.R. 3
Ayer's Cliff, Quebec
Canada JOB 1CO

Maximum, Inc.
42 South Ave.
Natick, MA 01760
(617) 785–0113

Meteorology Research, Inc.
P.O. Box 637
Altadena, CA 91001
(213) 791–1901

Metway Electrical Industries Ltd.
Canning Street
Brighton, Sussex
England

Millville Windmills & Solar Equipment Co.
Box 32–10335 Old 44 Drive
Millville, CA 96062

Mule Battery Company
325 Vallet Street
Providence, RI 02908
(401) 421–3773

Natural Energy Centre
2 York Street
London W.1.
England

Natural Energy Systems
Callen Avenue
Kent City, MI 49330

Natural Power, Inc.
New Boston, NH 03070
(603) 487–2456

Natural Power Systems, Inc.
3316 Augusta Avenue
Omaha, NB 68144

North Wind Power Co.
Box 315
Warren, VT 05674

O'Brock Windmill Sales
Route 1, 12th Street
North Benton, OH 44449

P.I. Enquiries Ltd.
The Dean
Alresford, Hants
England

Pacific Energy Systems
615 Romero Canyon Road
Santa Barbara, CA 93018

Penwalt Automatic Power
213 Hutcheson Street
Houston, TX 77003

Penwalt Corporation
P.O. Box 18738
Houston, TX 77023

Pinson Energy Corp.
Box 7
Marston Mills, MA 02648

Prairie Sun and Wind Co.
4408 62nd St.
Lubbock, TX 79409

Price Properties
5211 S.W. Vermont Street
Portland, OR 97219

Propeller Engineering Duplicating
403 Avenida Teresa
San Clemente, CA 92672
(714) 498–3739

Real Gas and Electric Co.
Box A
Guerneville, CA 65446

Rohn Tower
Division of Unarco Industries, Inc.
P.O. Box 2000
Peoria, IL 61601
(309) 697–4400

Sackett, William
Rt. 1, Box 289L
Gouldsboro, PA 18424

Sancken Wind Electric, Inc.
Kingman, AZ 86401

F. L. Schmidt Co.
69 Skt Klemensuzj
Hjallese, Denmark 5260

Schupbach, Ralph
321 13th Street
Alva, OK 73717

Sencenbaugh Wind Electric
P.O. Box 11174
Palo Alto, CA 94306

Senich Corp.
Box 1168
Lancaster, PA 17609

Shingletown Electric
P.O. Box 237
Shingletown, CA 96008

Sidney Williams & Co., Ltd.
P.O. Box 22
Dulwich Hill, N.S.W.
Australia 2203

Sign X Laboratories, Inc.
Essex, CT 06426
(203) 767–1700

R. A. Simerl Instrument Division
238 West St.
Annapolis, MD 21401
(301) 849–8667

Sine-Sync™ Model 130
Real Gas & Electric Co., Inc.
P.O. Box F
Santa Rosa, CA 95402

Solar Energy Company
810 18th St. N.W.
Washington, DC 20006

Solargy Corporation
17914 E. Warren Avenue
Detroit, MI 48224

Solar Wind Company
RFD 2
East Holden, ME 04429

Soleq Corporation
5969 Elston Avenue
Chicago, IL 60646
(312) 792–3811

Southern Cross Machinery Ltd.
P.O. Box 424
Toowomba, Queensland
Australia 4350

Southern Steel Works Ltd.
Ballyhale
Co. Kilkenny
Ireland

M. C. Stewart Co.
Ashburnham, MA 01430

Storage Battery Technical Service Manual
Battery Council International
111 East Wacker Drive
Chicago, IL 60601
(312) 644–6610

Sunflower Power Company
Route 1, Box 93–A
Oskaloosa, KS 66066

Sunglow Alternative Energy Systems
8261 West State Street
Boise, ID 83702

Surrette Storage Battery Co., Inc.
Box 3027
Salem, MA 01970
(617) 754–4444

Topaz Electronics
3855 Ruffin Road
San Diego, CA 92123

Trimble Windmills
Crimple Grange
Beckwithshaw
Harrogate, North Yorkshire
England

Trojan Batteries
1125 Mariposa Street
San Francisco, CA 94107
(415) 864–1565

United Technologies Research Ctr.
Silver Lane
East Hartford, CT 06108

U.S.S.R. Enercogomashexport
35 Mosfilmovskaya Ul
Moscow V 330
U.S.S.R.

Wadler Mfg. Co.
Galena, KS 66739

Wakes & Lamb Ltd.
Millgate Works
Newark, Notts
England

Westburg Manufacturing Co.
3400 Westach Way
Sonoma, CA 95476
(707) 938–2121

West Wind
Box 1465
Farmington, NM 87401

Windco Division
Dyna Technology Inc.
P.O. Box 3263
Sioux City, IA 51102

Wind Energy Services Co.
P.O. Box 458
70 Marston Ave.
Hyannis Port, MA 02647
(617) 775–3334

Wind Energy Supply Co. Ltd.
Bolney Avenue
Peacehaven, Sussex
England

Wind Engineering Corporation
Box 5926
Lubbock, TX 79417

Windependence Electric
P.O. Box M1188
Ann Arbor, MI 48106

Windflower Wind Computer
Lund Enterprises, Inc.
1180 Industrial Avenue
Escondido, CA 92025

Windlite
Box 43
Anchorage, AK 99510

Wind Power Systems, Inc.
P.O. Box 17323
San Diego, CA 92117
(714) 452–7040

Wind, Wood & Water
P.O. Box 193
Shokan, NY 12481

Windworks
Box 329, Route 3
Mukwonago, WI 53149

Windy Ten
Box 111
Shelby, MI 49445

W. T. G. Energy Systems
Box 87, 1 LaSalle
St. Angelo, NY 14006

Wyatt Brothers Ltd.
Whitchurch, Salop
England

Zephyr Wind Dynamo Co.
Box 241
21 Stanwood St.
Brunswick, ME 09011

Source/Resource

EQUIPMENT

Comments regarding selection of wind system components and the identification of common factors or specifications which make it possible to compare machines are presented through the courtesy of Michael Evans, editor and publisher of *Wind Power Digest,* published quarterly at Bristol, Indiana 46507.

SELECTING WIND SYSTEM COMPONENTS

There are four major kinds of wind energy system components: (1) the wind machine itself, which provides either mechanical or electrical energy; (2) the tower, which is the support structure for the wind machine; (3) the storage system, which in today's marketplace consists of either storage batteries or some form of interface with the utility power line; and (4) electrical subcomponents, such as inverters, voltage regulators, and automatic switching devices. Each Wind Energy Conversion System (WECS) manufacturer or distributor has a preferred package of components. Likewise, each manufacturer or distributor will offer many compelling reasons why particular products are best. The following is designed to help you ask the right questions when you shop for a wind energy system and to help you determine which system is best suited to your needs.

THE WIND MACHINE

A review of the equipment that follows quickly reveals an array of machines with remarkably dissimilar designs. Some have two blades, some three; some revolve around a horizontal axis like the farm windmill, and others revolve on a vertical axis like a merry-go-round. Each machine has been designed to meet certain operational specifications, and each has its own special features. Although many of the machines differ in many respects, it is possible to compare each machine against the next on many points. The following are those common factors or specifications. In cases where these specifications are somewhat technical in nature, an explanation is included.

Model—The machine model number specified by the manufacturer.

Rotor diameter—A measurement of the rotor or blades of a wind machine derived by measuring from the center of the rotation (the generator shaft) to the tip of the blade and then multiplying by two. This measurement helps determine the total area of wind flow swept by the blades.

System weight—Includes all the components of the wind machine found above the top of the tower support structure, and in some cases, part of the tower itself.

Rotor weight—Total weight of all blades, including the blade hub.

Cut-in wind speed—The wind speed at which the wind machine first produces a usable amount of power. This is an important specification because winds below the cut-in speed cannot be used to provide power. The cut-in wind speed should never be more than the average wind speed at the intended site.

Shut-down wind speed—Wind speed at which the wind machine ceases to operate. The blades are motionless either because of a lack of sufficient wind or because they are held motionless by a brake.

Rated output—The generating capacity of the wind machine as set by the manufacturer and the lowest wind speed at which the generating capacity is achieved. Generally the lower the wind speed at which the rated power occurs, the greater the total output of the wind machine.

Maximum output—The maximum generating capacity or output of a wind machine at the lowest wind at which the output occurs.

RPM at rated output—The number of revolutions of the blades per minute when the wind machine is producing its rated output.

Overspeed control—The design methods used to prevent the blades from turning too rapidly in high wind conditions.

Generator/Alternator—Describes the kind of generating component used in the wind machine and the kind of drive mechanism employed in the design. Direct drive is a direct link between the blade hub and the generator. Gear boxes step up the speed of the drive shaft before it reaches the generator.

Testing procedures

Warranty

Maintenance schedule

In addition to the above specifications, several other points should be investigated by the potential user. How many wind machines of any given type are up and in actual operation? What data is available from operational tests? Has a machine been tested at the federally-sponsored Rocky Flats Small Wind Machine System Testing Center in Colorado? Will the machine be tested at Rocky Flats in the near future? It is always best to ask as many questions and to uncover as much information as possible before making a purchase.

WIND POWER DIGEST is published quarterly in March, June, September, and December of each year by Michael Evans, 54468 CR 31, Bristol, IN 45607.

NOTE

Many of the manufacturers and distributors listed offer additional literature, specifications and models. Most make information available for free or a nominal charge.

Aermotor (Division of Valley Industries)

Industrial Park—P.O. Box 1364
Conway, AR 72032
(501) 329–9811

Contact: Stan Anderson, coordinator
Machine description: Up-wind, horizontal-axis, water pumpers.

Model: 702-6 ft.

Rotor diameter: 6 feet (1.82 meters)
Rotor weight: Not available
System weight: 210 lbs. (95.25 kg.)
Blade materials: Galvanized steel
Cut-in wind speed: 9 mph (14.48 kmph)
Shut-down wind speed: 28 mph (45.06 kmph)
Rated output: 15–20 mph (24.14–32.18 kmph)
Maximum output: 15–20 mph (24.14–32.18 kmph)
RPM at rated output: 125
Overspeed control: Rotor turns sideways to wind
Testing procedures: 44 years of manufacturing
Warranty: One year, materials and workmanship
Maintenance schedule: Annual lubrication

Model: 702-8 ft.

Rotor diameter: 8 feet (2.43 meters)
Rotor weight: Not available
System weight: 355 lbs. (161 kg.)
Blade materials: Galvanized steel
Cut-in wind speed: 9 mph (14.48 kmph)
Shut-down wind speed: 28 mph (45.06 kmph)
Rated output: 15–20 mph (24.14–32.18 kmph)
Maximum output: 15–20 mph (24.14–32.18 kmph)
RPM at rated output: 105
Overspeed control: Rotor turns sideways to wind
Testing procedures: 44 years of manufacturing
Warranty: One year, materials and workmanship
Maintenance schedule: Annual lubrication

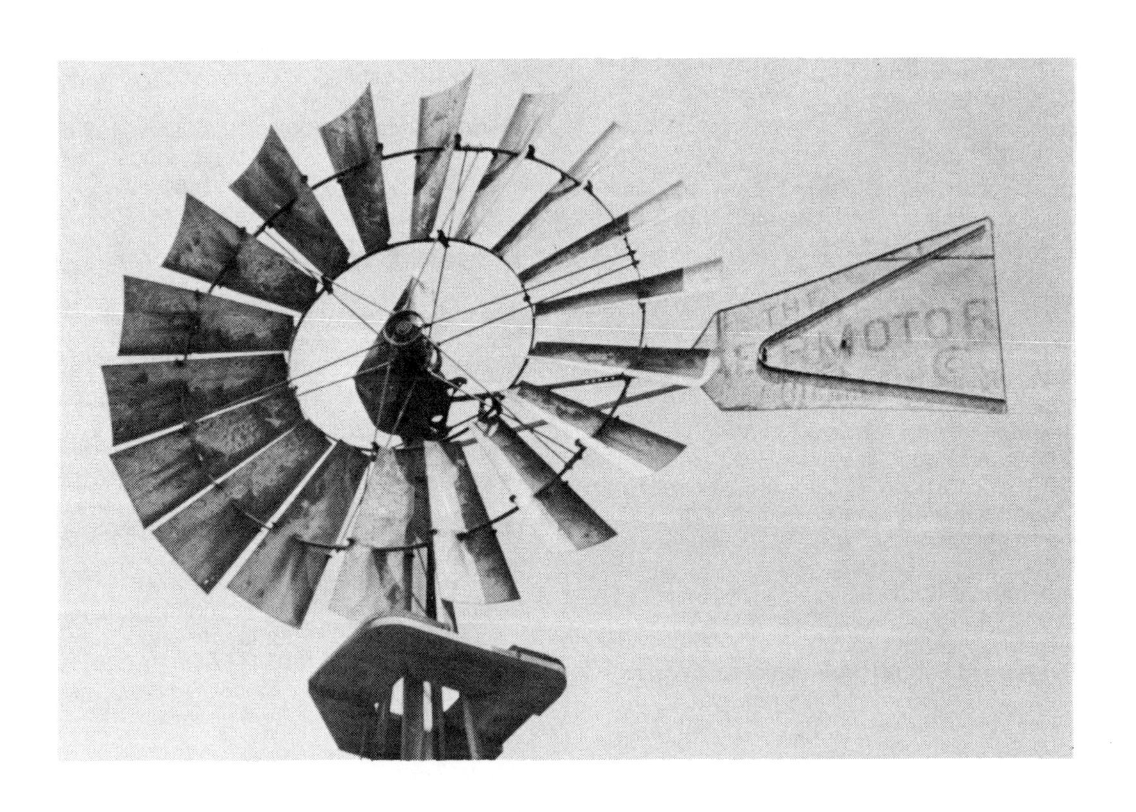

Aero Power Systems Inc.

2398 Fourth Street
Berkeley, CA 97410
(415) 848–2710
Contact: Mario Agnello
Machine description: Up-wind, horizontal-axis, three blades.

Model: SL 1500

Rotor diameter: 10 feet (3.04 meters)
Rotor weight: 50 lbs. (22.67 kg.)
System weight: 160 lbs. (72.57 kg.)
Blade materials: Wood, sitka spruce
Cut-in wind speed: 6 mph (9.65 kmph)
Shut-down wind speed: 100 mph (160.93 kmph)
Rated output: 1,430 watts at 25 mph (40.23 kmph)
Maximum output: 1,600 watts at 30 mph (48.28 kmph)
RPM at rated output: 500
Overspeed control: Mechanical, variable pitch, centrifugally activated.
Generator/Alternator: 14 or 18 VAC, 3 phase, DC output
Testing procedures: Field operation
Warranty: One year, defects in workmanship and materials
Maintenance schedule: Semi-annual, grease hub, check blades

Altos—The Alternate Current

P.O. Box 905
Boulder, CO 80302
(303) 442–0855
Contact: Michael Blakley or Edward Gitilin
Machine description: Up-wind, horizontal-axis, multibladed

Model: BWP-8B

Rotor diameter: 7.6 feet (2.31 meters)
Rotor weight: 59 lbs. (26.76 kg.)
System weight: 250 lbs. (113.39 kg.)
Blade materials: Aluminum, 5052
Cut-in wind speed: 10 mph (16.09 kmph)
Shut-down wind speed: 75 mph (120.7 kmph)
Rated output: 1,500 watts at 28 mph (45.06 kmph)
Maximum output: 2,000 watts at 40 mph (64.37 kmph)
RPM at rated output: 165
Overspeed control: Aerodynamic drag, rotor turns sideways
Generator/Alternator: 24 VDC, 0–70 amp
Testing procedures: Field testing, free steam anemometer
Warranty components: One year, parts and workmanship
Warranty turbine: 90 days, parts and workmanship
Maintenance schedule: Semi-annual fluid and system check, annual inspection

American Wind Turbine Co.

1016 E. Airport Road
Stillwater, OK 74074
(405) 377–5333
Contact: Nancy Thedfor, office manager
Machine description: Up-wind, horizontal-axis, multibladed.

Model: 12 ft.

Rotor diameter: 11.5 feet (3.5 meters)
Rotor weight: 92 lbs. (41.73 kg.)
System weight: 320 lbs. (145.14 kg.)
Blade materials: Aluminum
Cut-in wind speed: 10 mph (16.09 kmph)
Shut-down wind speed: 35 mph (56.32 kmph)
Maximum output: Not available
RPM at rated output: 120
Overspeed control: Mechanical, tail turns rotor
Generator/Alternator: Permanent magnet, three phase alternator
Testing procedures: Moving test bed, prony brake
Warranty: Limited 90 days, parts and workmanship
Maintenance schedule: Semi-annual; check belts, annual grease and inspect

Model: BWP-12A

Rotor diameter: 11.5 feet (3.53 meters)
Rotor weight: 111 lbs. (50.06 kg.)
System weight: 300 lbs. (136.07 kg.)
Blade materials: Aluminum, 6061
Cut-in wind speed: 8 mph (12.87 kmph)
Shut-down wind speed: 60 mph (96.56 kmph)
Rated output: 2200 watts at 28 mph (45.06 kmph)
Maximum output: 4000 watts at 40 mph (64.37 kmph)
RPM at Rated output: 116
Overspeed control: Aerodynamic drag, rotor turns sideways
Generator/Alternator: 115/200 VAC, 3 phase
Testing procedures: Calculated data. Tests now underway

Warranty components: One year, parts and workmanship
Warranty turbine: 90 days parts and workmanship
Maintenance schedule: Semi-annual fluid and system check and annual inspection

Carter Motor Company

2711 West George Street
Chicago, IL 60618
(312) 588–7700
Machine description: DC to AC converters

	DC Input		115V. AC
Code No.	Volts	Amps	Watts
K10100M	230	6	1000
K10150M	230	10	1500
K10200M	230	12	2000
D10100M	115	13	1000
D10150M	115	20	1500
D10200M	115	24	2000
W10100M	48	33	1000
W10150M	48	48	1500
W10200M	48	60	2000
C10100M	32	50	1000
C10150M	32	65	1500
C10200M	32	90	2000
J10100M	28	52	1000
J10150M	28	80	1500
JS10200M*	28	103	2000
E10100M	24	68	1000
E10150M	24	100	1500

All models are for 115 AC 60 HZ single phase output.
Continuous duty at 100% power factor.
*1 hour duty cycle.

	DC Input		115V. AC
Code No.	Volts	Amps	Watts
DR1021CP	115	2.5	210
DR1050CP	115	7.0	500
JR1021CP	28	14.0	210
JR1030CP	28	20	300
JR1050CP	28	28	500
ER1021CP	24	16	210
ER1030CP	24	22	300
ER1050CP	24	33	500
BR1021CP	12	29	210
BR1030CP	12	45	300
BRS1040CP*	12	56	400

*One hour duty cycle.

Climet Instruments Company

P.O. Box 151
1351 West Colton Avenue
Redlands, CA 92373
(714) 793–2788
Machine description: Environmental monitoring systems

Model: 012-15 Wind Direction Transmitter

This very precise research-grade sensor has a threshold of .75 mph (0.34 mps) (1.206 kmph; 2 kmps). A damping ratio of 0.4 or 0.6 allows fast rise time with minimum overshoot.

The 012-15 is comprised of a symmetrical airfoil vane and integral drive shaft, a precision potentiometer assembly, and the housing with related fittings.

Selection of the appropriate printed circuit board provides the option of a 0–360° or 0–540° range.

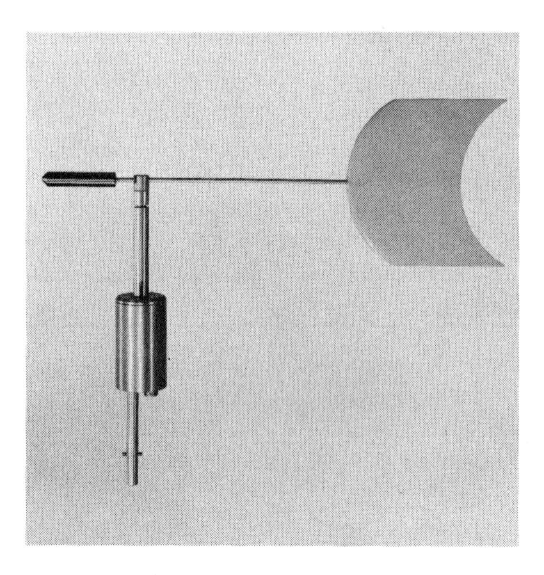

Model: 018-10 (540°)
Model: 018-30 (360°)
Axiometer Wind Transmitters

Featuring the latest concepts in axial-flow anemometry, these research-grade sensors are designed for the simultaneous measurement of horizontal and vertical wind direction, as well as axial wind speed. They respond to winds as low as 0.5 mph (0.8 kmph) and have a damping ratio of 0.6 to allow a fast rise time with minimum overshoot.

The lightweight propeller provides a distance constant of 3 feet (0.91 meters) with a starting threshold of less than 0.5 mph (0.22 mps)(0.8 kmph; 0.0013 kmphs).

Both transmitters are high-reliability instruments of lightweight but extremely rugged design and operate in almost any environment in the temperature range of −50°F to +155°F (−45.6°C to 68.3°C) over extended periods of time.

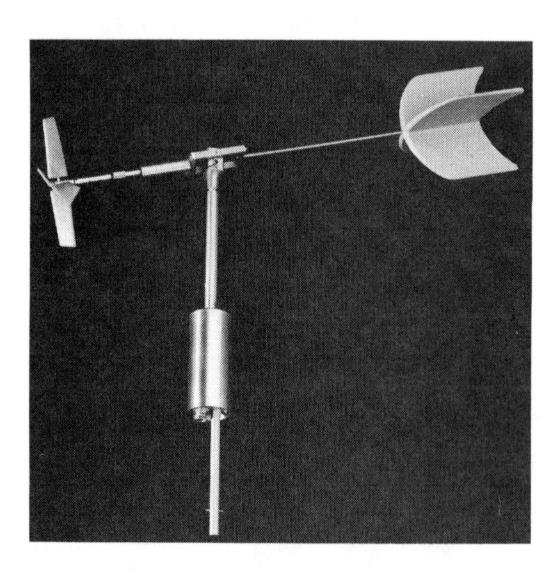

Dakota Wind and Sun Ltd.

P.O. Box 178
Aberdeen, SD 57401
(605) 229–0815
Contact: Orville Lynner or Paul Biorn
Machine description: Up-wind, horizontal-axis, three blades

Model: BC4

Rotor diameter: 14 feet (4.26 meters)
Rotor weight: 75 lbs. (34.01 kg.)
System weight: 600 lbs. (272.15 kg.)
Blade materials: Wood (Sitka spruce), epoxy paint
Cut-in wind speed: 8 mph (12.87 kmph)
Shut-down wind speed: 40 mph (64.37 kmph)
Rated output: 4,000 watts at 27 mph (43.45 kmph)
Maximum output: 4,000 watts at 27 mph (43.45 kmph)

Overspeed control: Centrifugal blade feathering

Generator/Alternator: 110 volt direct current generator

Testing procedures: Field tests on 50-foot (15.24-meter) tower

Warranty: Limited, replace defective parts for two years

Maintenance schedule: Semi-annual; inspect, lubricate system

Delatron Systems Corporation

553 Lively Boulevard
Elk Grove Village, IL 60007

Delatron 3 KVA, Type VC inverter with removable electronics chassis and plug-in circuit boards. This inverter has been specifically designed for use in wind power systems and small hydroelectric systems.

Delatron batteries designed for deep cycle applications such as wind power and hydroelectric systems.

Dempster Industries, Inc.

P.O. Box 848
Beatrice, NB 68310
(402) 223–4026
Contact: Sales department
Machine description: Up-wind, horizontal axis, water-pumpers

Model: 6 ft.

Rotor diameter: 6 feet (1.82 meters)
Rotor weight: 100 lbs. (45.35 kg.)
System weight: 280 lbs. (127 kg.)
Blade materials: Galvanized steel
Cut-in wind speed: 5 mph (8.04 kmph)
Shut-down wind speed: 50 mph (80.46 kmph)
Rated output: See chart, 15 mph (29.14 kmph)
RPM at rated output: Not available
Overspeed control: Rotor turns sideways to the wind
Testing procedures: Data calculated and tested
Warranty: Limited five years, parts and workmanship
Maintenance schedule: Annual inspection and lubrication

Model: 8 ft.

Rotor diameter: 8 feet (2.43 meters)
Rotor weight: 120 lbs. (54.43 kg.)
System weight: 388 lbs. (176 kg.)
Blade materials: Galvanized steel
Cut-in wind speed: 5 mph (8.04 kmph)
Shut-down wind speed: 50 mph (80.46 kmph)
Rated output: See chart, 15 mph (29.14 kmph)
RPM at rated output: Not available
Overspeed control: Rotor turns sideways to wind
Testing procedures: Calculated and tested
Warranty: Limited five years, parts and workmanship
Maintenance schedule: Annual inspection and lubrication

Dunlite Electrical Products

c/o Enertech Corp.
P.O. Box 420
Norwich, VT 05055
(802) 649–1145
Contact: Ned Coffin or Robert Sherwin
Machine description: Up-wind, horizontal-axis, three blades

Standard Model (81–002550)

Rotor diameter: 13.5 feet (4.11 meters)
Rotor weight: 130 lbs. (58.96 kg.)
System weight: 500 lbs. (226.79 kg.)
Blade materials: Galvanized sheet steel
Cut-in wind speed: 8 mph (12.87 kmph)
Shut-down wind speed: None
Rated output: 2,000 watts at 25 mph (40.23 kmph)
Maximum output: 3,000 watts at 32 mph (51.49 kmph)
RPM at rated output: 185
Overspeed control: Mechanical, centrifugal blade feathering
Generator/Alternator: 12/24/32/38/110 DVC three-phase alternator
Testing procedures: Wind tunnel tests
Warranty: One year, parts and workmanship
Maintenance schedule: Annual; check oil and inspect system

High Wind Speed Model

Rotor diameter: 10 feet (3.04 meters)
Rotor weight: 130 lbs. (58.96 kg.)
System weight: 500 lbs. (226.79 kg.)
Blade materials: Galvanized sheet steel
Cut-in wind speed: 14 mph (22.53 kmph)
Shut-down wind speed: None
Rated output: 1,000 watts at 37 mph (59.54 kmph)
Maximum output: 1,200 watts at 80 mph (128.74 kmph)
RPM at rated output: Not available
Overspeed control: Centrifugally activated blade feathering
Generator/Alternator: 12/24/32/110 volt alternator
Testing procedures: Wind tunnel
Warranty: One year parts and workmanship

Dynergy Corporation

P.O. Box 428
1269 Union Avenue
Laconia, NH 03246
(603) 524–8313
Contact: Robert B. Allen, vice president
Machine description: Vertical-axis, three blades

Performance and evaluation tests on 110V, 2 KW Dunlite wind generator at Flinders University, Adelaide, Australia.

Model: 5 Meter

Rotor diameter: 15 feet (4.57 meters)
Rotor weight: 312 lbs. (141.52 kg.)
System weight: 850 lbs. (385.55 kg.)
Blade materials: Aluminum 6061-T6
Cut-in wind speed: 10 mph (16.09 kmph)
Shut-down wind speed: Not available
Rated output: 3,300 watts at 24 mph (38.62 kmph)
Maximum output: Not available
RPM at rated output: 200
Overspeed control: Disc brake, aerodynamic stall
Testing procedures: Test runs
Warranty: Available
Maintenance schedule: Semi-annual inspection of blades

Energy Development Co.

179E R.D. 2
Hamburg, PA 19526
(215) 562–8856
Contact: Terrance Mehrkam
Machine description: Down-wind, horizontal axis, four blades

Model: 440

Rotor diameter: 38 feet (11.58 meters)
Rotor weight: 925 lbs. (419.58 kg.)
System weight: 7,000 lbs. (3,175.14 kg.)
Blade materials: Aluminum T-6
Cut-in wind speed: 5 mph (8.04 kmph)
Shut-down wind speed: 40 mph (64.37 kmph)
Rated output: 20,000 watts at 25 mph (40.23 kmph)
Maximum output: 20,000 watts
RPM at rated output: 60
Overspeed control: Mechanical brake
Generator/Alternator: Synchronous, 3 phase, 240 VAC
Testing procedures: Monitored during operation
Warranty: Two years, materials and workmanship
Maintenance schedule: Lubricate every six months

Enertech Corporation

P.O. Box 420
Norwich, VT 05055
(802) 649–1145
Contact: Ned Coffin or Bill Drake
Machine description: Down-wind, horizontal-axis, three blades

Heller-Aller Company

Perry and Oakwood Streets
Napoleon, OH 43545
Contact: James Bradner, vice president
Machine description: Up-wind, horizontal-axis, water-pumpers

Model: Baker 10 ft.

Rotor diameter: 10 feet (3.04 meters)
Rotor weight: Not available
System weight: 475 lbs. (215.46 kg.)
Blade materials: Galvanized steel
Cut-in wind speed: 7 mph (11.26 kmph)
Shut-down wind speed: 25 mph (40.23 kmph)
Rated output: See chart, 15 mph (24.14 kmph)
RPM at rated output: 150
Overspeed control: Rotor turns sideways to the wind
Testing procedures: Not available
Warranty: One year, parts and workmanship
Maintenance schedule: Annual inspection and lubrication

Model: 1500

Rotor diameter: 13.2 feet (4.02 meters)
Rotor weight: 48 lbs. (77.24 kg.)
System weight: 185 lbs. (83.91 kg.)
Blade materials: Wood
Cut-in wind speed: 9 mph (14.48 kmph)
Shut-down wind speed: 40 mph (64.37 kmph)
Rated output: 1,500 watts at 22 mph (35.4 kmph)
Maximum output: 1,650 watts at 25 mph (40.23 kmph)
RPM at rated output: 170
Overspeed control: Automatic brake
Generator/Alternator: Induction generator, 115 VAC
Testing procedures: Truck, field testing
Warranty: One year, parts and workmanship
Maintenance schedule: Annual oil change and system inspection

Model: Baker 6 ft.

Rotor diameter: 6 feet (1.82 meters)
Rotor weight: Not available
System weight: 220 lbs. (99.76 kg.)
Blade materials: Galvanized steel
Cut-in wind speed: 7 mph (11.26 kmph)
Shut-down wind speed: 25 mph (40.23 kmph)
Rated output: See chart, 15 mph (24.14 kmph)
RPM at rated output: 150

Overspeed control: Rotor turns sideways to the wind
Testing procedures: Not available
Warranty: One year, parts and workmanship
Maintenance schedule: Annual inspection and lubrication

Model: Baker 12 ft.

Rotor diameter: 12 feet (3.65 meters)
Rotor weight: Not available
System weight: 800 lbs. (362.87 kg.)
Blade materials: Galvanized steel
Cut-in wind speed: 7 mph (11.26 kmph)
Shut-down wind speed: 25 mph (40.23 kmph)
Rated output: See chart, 15 mph (24.14 kmph)
RPM at rated output: 150
Overspeed control: Rotor turns sideways to the wind
Testing procedures: Not available
Warranty: One year, parts and workmanship
Maintenance schedule: Annual inspection and lubrication

Independent Energy Systems, Inc.

6043 Sterrettania Rd.
Fairview, PA 16415
(814) 833–3567
Contact: John D'Angelo, president
Machine description: Up-wind, horizontal-axis, three blades.

Model: Sky Hawks I & II

Rotor diameter: 15.66 feet (4.77 meters)
Rotor weight: 35 lbs. (15.87 kg.)
System weight: 600 lbs. (272.15 kg.)
Blade materials: Aircraft quality Sitka spruce
Cut-in wind speed: 7–8 mph (11.26–12.87 kmph)
Shut-down wind speed: 40 mph (64.37 kmph)
Rated output: 4,000 watts at 23 mph (37.01 kmph)
Maximum output: 4,500 watts at 30 mph (48.28 kmph)
RPM at rated output: 285
Overspeed control: Centrifugally activated blade pitching
Generator/Alternator: Generator
Testing procedures: Bench testing, on site testing
Warranty: Two years, materials and workmanship
Maintenance schedule: Lubrication of 5 zurk fittings twice a year

Kedco, Inc.

9016 Aviation Blvd.
Inglewood, CA 90301
(213) 776–6636
Contact: Terry Rainey, wind program manager
Machine description: Down-wind; horizontal-axis, three blades

Model: 1200

Rotor diameter: 12 feet (3.65 meters)
Rotor weight: 71 lbs. (32.2 kg.)
System weight: 202 lbs. (91.62 kg.)

Blade materials: Aluminum, 2024-T3
Cut-in wind speed: 7 mph (11.26 kmph)
Shut-down wind speed: 70 mph (112.65 kmph)
Rated output: 1,200 watts at 22 mph (35.4 kmph)
Maximum output: 1,200 watts at 22 mph (35.4 kmph)
RPM at rated output: 300
Overspeed control: Mechanical: centrifugal blade feathering
Generator/Alternator: 14.4 VAC
Testing procedures: Moving test bed
Warranty: One year, parts and labor
Maintenance schedule: Annual: check fluid levels, inspect system

Millville Wind & Solar Equipment Co.

P.O. Box 32—10335 Old Drive, Millville, CA 96062
(916) 547–4302
Contact: Devon Tassen, president
Machine description: Up-wind, horizontal-axis, three blades.

Model: 10-3-Ind

Rotor diameter: 25 feet (7.62 meters)
Rotor weight: 230 lbs. (104.32 kg.)
System weight: 850 lbs. (385.55 kg.)
Blade materials: Aluminum
Cut-in wind speed: 9 mph (14.48 kmph)
Shut-down wind speed: 60 mph (95.56 kmph)
Rated output: 10,000 watts at 25 mph (40.23 kmph)
Maximum output: 10,000 watts at 25 mph (40.23 kmph)
RPM at rated output: 80
Overspeed control: Mechanical; blade feathering
Generator/Alternator: 220V, inducting generator
Testing procedures: Field testing
Warranty: One year
Maintenance schedule: Semi-annual, oil check and system inspection

North Wind Power Co.

P.O. Box 315
Warren, VT 05674
(802) 496–2995
Contact: Don Mayer, president
Machine description: Up-wind, horizontal-axis, three blades.

Model 2Kw, 32V

Rotor diameter: 13.6 feet (4.14 meters)
Rotor weight: 70 lbs. (31.75 kg.)
System weight: 480 lbs. (217.72 kg.)
Blade materials: Wood (Spitka spruce), fiberglass coating
Cut-in wind speed: 8 mph (12.87 kmph)
Shut-down wind speed: None
Rated output: 2,000 watts at 22 mph (35.4 kmph)
Maximum output: 3,000 watts at 90 mph (144.48 kmph)
RPM at rated output: 265
Overspeed control: Mechanical, centrifugal blade pitching
Generator/Alternator: 32 VDC generator—direct drive
Testing procedures: Dynamometry, voltmeter, ammeter
Maintenance schedule: Grease every three to five years—clean or replace brushes, refinish blades every five years.
Warranty: One year unconditional, generator and main components.

A true believer! A 5-story Dutch-style windmill which has a 44-foot diameter rotor and drives a 30 KW generator. (Devon Tassen, President, Millville Windmills and Solar Equipment Co.)

Pennwalt Automatic Power

P.O. Box 18738
Houston, TX 77023
(713) 228–5208

Pinson Energy Corporation

P.O. Box 7
Marston Mills, MA 02648
(617) 428–8535
Contact: Herman Drees, president
Machine description: Vertical-axis, three blades.

Model: C2E

Rotor diameter: 12 feet (3.65 meters)
Rotor weight: 234 lbs. (106.14 kg.)
System weight: 439 lbs. (199.13 kg.)
Blade materials: Aluminum 6061-T6
Cut-in wind speed: 7 mph (11.26 kmph)
Shut-down wind speed: 40 mph (64.37 kmph)
Rated output: 2,000 watts at 24 mph (38.62 kmph)
Maximum output: 4,000 watts at 30 mph (48.28 kmph)
RPM at rated output: 160
Overspeed control: Mechanical, centrifugal blade pitching
Generator/Alternator: 120V-33A or 240-V-16 internal excitation
Testing procedures: Tachometer, anemometer, voltmeter, resistive load
Warranty: One year, all parts and service
Maintenance schedule: Annual lubrication of bearings, tighten hardware, replace drive belt

Sencenbaugh Wind Electric

P.O. Box 1174
Palo Alto, CA 94306
(415) 964–1593
Contact: Jim Sencenbaugh, president
Machine description: Up-wind, horizontal-axis, three blades.

Model: 400–14 HDS

Rotor diameter: 7 feet (2.13 meters)
Rotor weight: 3 lbs. (1.36 kg.)
System weight: 65 lbs. (29.48 kg.)
Blade materials: Wood (Sitka spruce) polyurethane finish
Cut-in wind speed: 9 mph (14.48 kmph)
Shut-down wind speed: 60 mph (96.56 kmph)
Rated output: 400 watts at 20 mph (32.18 kmph)
Maximum output: 500 watts at 26 mph (41.84 kmph)
RPM at rated output: 1,000
Overspeed control: Rotor tilts upward in excessive winds
Generator/Alternator: 14VDC, self-excitation
Testing procedures: Moving test bed
Warranty: One year, parts and labor
Maintenance schedule: Annual inspection

R. A. Simerl

238 West Street
Annapolis, MD 21401

Simerl anemometer, Model RK-7KK, popular in overseas countries. Available with cup-type rotor shown at top of next page.

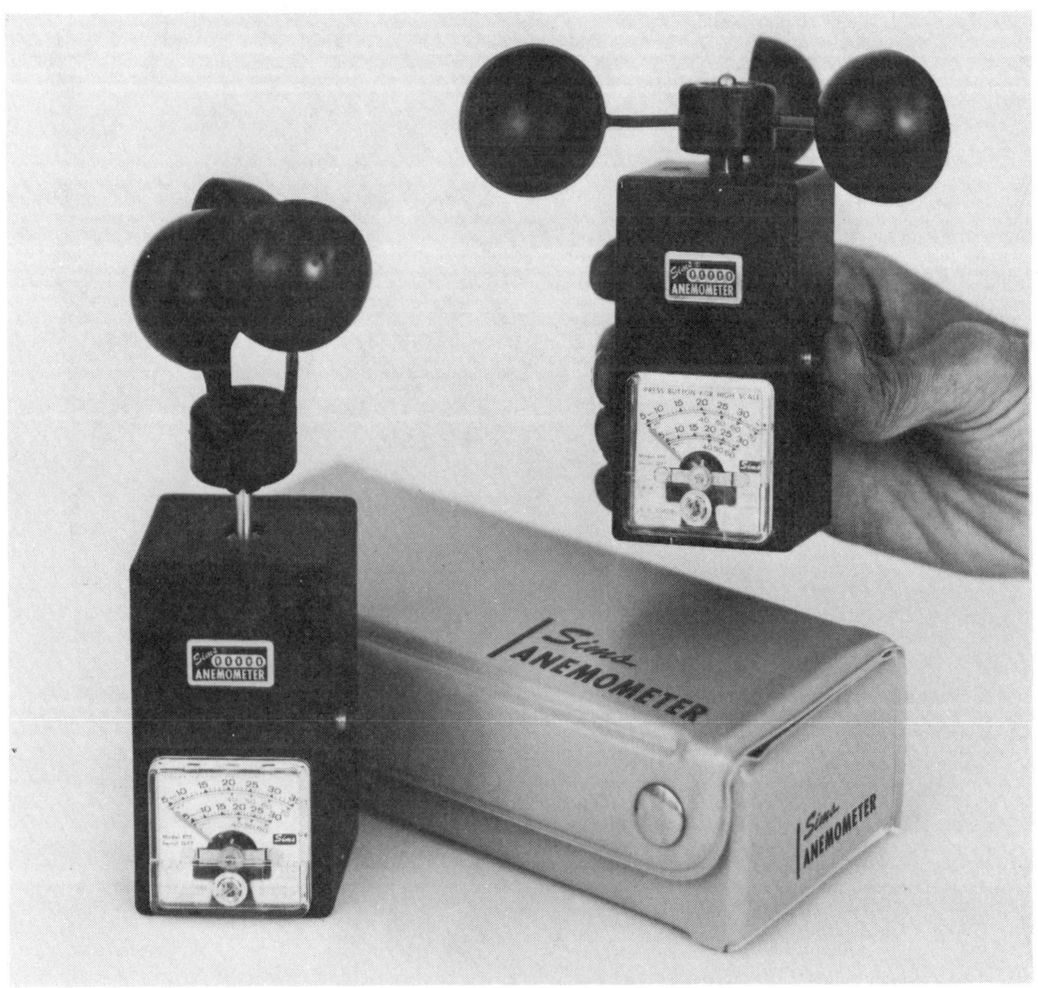

Simerl Model BTC anemometers. Just hold in the wind and read wind velocity direct from scale. Shown with cups in folded and in operating modes.

Soleq Corporation

5969 Elston Ave.
Chicago, IL 60646
(312) 792–3811

Soleq windmill inverter is a transistorized unit featuring sine wave regulated performance.

Windmill inverter sine wave regulated

Output: Sine wave, Voltage regulated ± 3%; Distortion approx. 5%.
Frequency: 60 HZ ± 0.01% absolute accuracy.
Overload: 100% inrush over rated capacity; (max frequency, 2 min. intervals).
Output short circuit: Completely protected against dead shorts in start or operating mode.

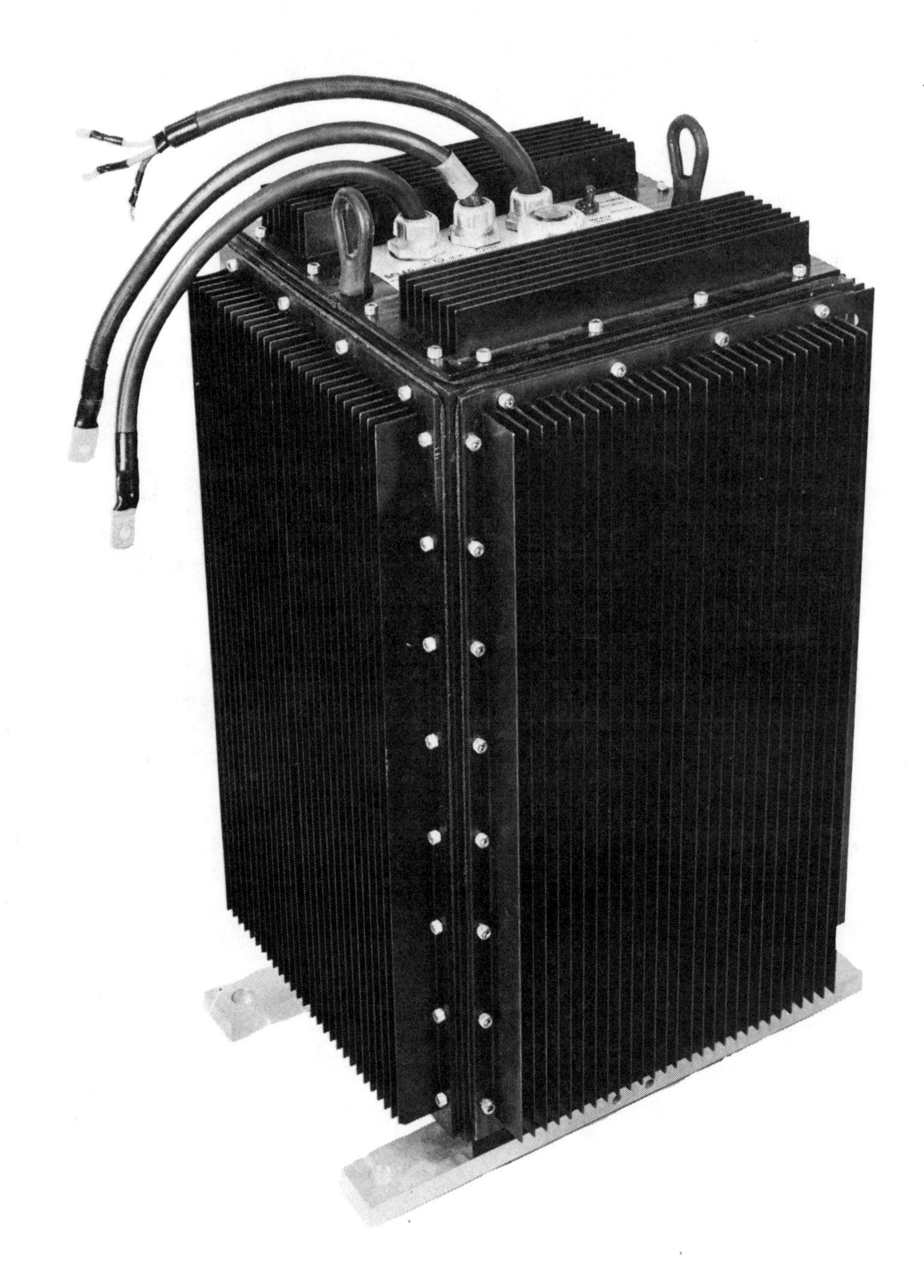

Short circuit input current: Limited to approximately 300% of rated current without damage to wiring, battery, or inverter.
No load power drain: 10 to 40 VA; proportional to input voltage. Units without load-sensor drain approx. 10% of rated capacity.
Efficiency: Approximately 90% at rated capacity.
Polarity: Input protected against reverse polarity.
Automatic restart: No reset required when inverter shuts down because high or low voltage supply limit was exceeded.
Control switch: Three position: ON (Manual override)—OFF Oper. temperature: –30°C to +50°C (–22.0°F to +122.0°F) ambient without derating.
Storage temperature: –30°C to +85°C. (–22.0°F to +185.0°F)
Diagnostic indicators: Cause of service interruption is indicated on the Diagnostic Indicator Dial assembly.
Construction: Fabricated finned aluminum housing.
Electrical connection: Terminated pig tails with compression terminals 18″ long.
Mounting: Removable bars with ⅞″ (22 mm) holes located 10½″ x 20″ (26.6 x 45.7 cm) for 6KVA and 8½″ x 17″ (21.5 x 43.1 cm) for 1½ and 3 KVA units.

Sine wave inverter vented housing

Circuitry: Transistorized with solid state control circuitry.
Output: Voltage regulated ± 3%. 240V C.T. (2x120V). Total output may be loaded on one side.
Frequency: 60 HZ. ± 0.01% accuracy quartz crystal controlled.
Wave form: Sine wave; distortion approximately 5%. Perfect symmetry.
Insulation: 1500V on AC and 500V on DC. Primary, secondary and optional circuits are isolated from each other and from ground.
Overload current capability: 100% of low supply voltage rating for two seconds with approximately a 15% voltage drop.
Efficiency: Approximately 90% at nominal voltage and capacity. Efficiency is somewhat lower for low capacity and or low voltage units.
No load drain w/o load sensor: Approximately 10% of low supply capacity.
Load sensor drain: 10 to 20 VA depending on system input voltage.
Output short circuit: Completely protected against dead shorts in start or operating mode.
Automatic restart: No reset required when inverter shuts down because high or low input voltage limits were exceeded.
Amplitude sensing: Shuts inverter down under a prolonged overload such as a locked compressor motor. Reset by a two (2) second interruption of the DC supply or the control switch.
Polarity: Input protected against reverse polarity.
Control switch: Front or rear panel mounted; also serves as the reset switch.
Remote control: Fused terminals are provided for a remote SPST switch for on/off and reset functions.
Junction box: Input, output and control terminals are provided in rear panel junction box.
Operating temperature: –40°C (–40°F) when humidity protected) to +50°C (+122°F) without derating.
Storage temperature: –40°C to +85°C (–40°F to +185°F).
Available options: AC Load Sensor or 50VA first stage; Diagnostic Indicator Dial; current sensor capacity limiting circuit; front or rear panel mounting for control switch; and Diagnostic Indicator Dial.
Construction: Conventional 17″ (43.1 cm) wide sheet metal housing with base mounting. Rack panel mounting angles supplied when specified.

Sparco (Denmark)

c/o Enertech
P.O. Box 420
Norwich, VT 05055
(802) 649–1145
Contact: Edmund Coffin or Robert Sherwin
Machine description: Up-wind, horizontal-axis, waterpumper

Model: Sparco

Rotor diameter: 4.17 feet (1.27 meters)
Rotor weight: 8 lbs. (3.62 kg.)
System weight: 58 lbs. (26.30 kg.)
Blade materials: Cast aluminum
Cut-in wind speed: 5 mph (8.04 kmph)
Shut-down wind speed: None
Rated output: Not rated
Maximum output: 58 gallons/hour (219.55 liters/hour)
RPM at rated output: Not available
Overspeed control: Mechanical, blade feathering
Warranty: One year, parts and workmanship
Maintenance schedule: Semi annual inspection, lubrication

Winco (Division of Dyna Technology)

7850 Metro Parkway
Minneapolis, MN 55420
(612) 853–8400
Contact: Sales department
Machine description: Up-wind, horizontal-axis, two blades

Model: 1222H

Rotor diameter: 6 feet (1.82 meters)
Rotor weight: 20 lbs. (9.07 kg.)
System weight: 134 lbs. (60.78 kg.)
Blade materials: Wood
Cut-in wind speed: 7 mph (11.26 kmph)
Shut-down wind speed: 70 mph (112.65 kmph)
Rated output: 200 watts at 23 mph (37.01 kmph)
Maximum output: 200 watts at 23 mph (37.01 kmph)
RPM at rated output: 900
Overspeed control: Air brake, centrifugally activated
Generator/Alternator: 12 VDC generator
Testing procedures: During operation for 43 years
Warranty: One year, parts and labor
Maintenance schedule: Annually, oil and inspect

Wadler Manufacturing Co., Inc.

Rt. 2, Box 76
Galena, KS 66739
(316) 783–1355
Contact: Jerry Wade
Machine description: Two-bladed, vertical-axis. Savonious rotor

Model: Wadler

Rotor diameter: Unavailable
Rotor weight: Unavailable
System weight: 16 lbs. (7.25 kg.)
Blade materials: Aluminum
Cut-in wind speed: 2 mph (3.22 kmph)
Shut-down wind speed: Not applicable
Rated output: Not applicable
Maximum output: Unavailable
RPM at rated output: Not applicable
Overspeed control: None

Whirlwind Power Company

Box 18530
Denver, CO 80218
(303) 534–1567
Contact: Elliott Bayly
Machine description: Down-wind; horizontal-axis, two blades.

Model: A

Rotor diameter: 10 feet (3.05 meters)
Rotor weight: Not available
System weight: 71 lbs. (32.21 kg.)
Blade materials: Wood (Sitka spruce)
Cut-in wind speed: 8 mph (12.88 kmph)
Shut-down wind speed: 50 mph (80.47 kmph)
Rated output: 2,000 watts at 25 mph (40.23 kmph)
Maximum output: 2,000 watts at 25 mph (40.23 kmph)
RPM at rated output: 900
Overspeed control: Electro-mechanical brake
Generator/Alternator: 240VAC permanent magnet
Testing procedures: Moving test bed
Warranty: One year, parts and labor
Maintenance schedule: Every five years: lubricate bearings, refinish blades

Wind Power Systems, Inc.

P.O. Box 17323
San Olego, CA 92117
(714) 452–7040
Contact: Ed Salter
Machine description: Down-wind; horizontal-axis, three blades.

Storm Master 10

Rotor diameter: 32.8 feet (10 meters)
Rotor weight: 285 lbs. (129.27 kg)
System weight: 875 lbs. (397 kg.)
Blade materials: Fiberglass shell, foam core
Cut-in wind speed: 8 mph (12.88 kmph)
Shut-down wind speed: 150 mph (241.40 kmph)
Rated output: 6,000 watts at 18 mph (28.97 kmph)
Maximum output: 6,000 watts
RPM at rated output: 130
Overspeed control: Blade stall, brake
Generator/Alternator: Variety available, including permanent magnet
Testing procedures: Data calculated
Warranty: One year, materials and workmanship
Maintenance schedule: Not available

Wind Wizard
(Aero Lectric Company)

13517 Winters Avenue
Cresaptown, MD 21502
(301) 729–2325
Contact: L. Michael Glick
Machine description: Up-wind, horizontal-axis, three blades.

Model: C9D

Rotor diameter: 9 feet (2.74)
Rotor weight: 22 lbs. (9.97 kg.)
System weight: 50 lbs. (22.67 kg.)
Blade materials: Wood, urethane, or fiberglass and composite
Cut-in wind speed: 9 mph (14.48 kmph)
Shut-down wind speed: 40 mph (64.37 kmph)
Rated output: 600 watts at 26 mph (41.84 kmph)
Maximum output: 770 watts at 30 mph (48.28 kmph)
RPM at rated output: 337 at 27 mph (43.45 kmph)
Overspeed control: Rotor turns sideways
Generator/Alternator: Alternator adjustable, voltage 12 to 18 volts
Testing procedures: Prony brake
Warranty: One year, parts and workmanship limited
Maintenance schedule: Semi-annual inspection of system

WTG Energy Systems

Box 87, 1 La Salle St.
Angola, NY 14006

Contact: Alfred J. Gross, director of marketing
Machine description: Up-wind, horizontal-axis, three blades

Model: MP1-200

Rotor diameter: 80 feet (24.38 meters)
Rotor weight: 15,000 lbs. (6,803.88 kg.)
System weight: 85,000 lbs.
Blade materials: Steel, steel tubing, galvanized steel
Cut-in wind speed: 8 mph (12.87 kmph)
Shut-down wind speed: 50 mph (80.46 kmph)
Rated output: 200,000 watts
Maximum output: 200,000 watts
Generator/Alternator: Synchronous generator
Testing procedures: During operation

Zephyr Wind Dynamo Company

P.O. Box 241, 21 Stanwood St.
Brunswick, ME 04011
(207) 725–6534
Contact: Willard Gillette, president
Machine description: Vertical-axis, gyromill.

Model: Tetrahelix S

Rotor diameter: 2 feet (0.6 meters)
Rotor weight: 3 lbs. (1.36 kg.)
System weight: 5 lbs. (2.26 kg.)
Blade materials: Dacron, nylon, aluminum, kevlar ties
Cut-in wind speed: 12 mph (19.31 kmph)
Shut-down wind speed: None
Rated output: 7 watts at 25 mph (40.23 kmph)
Maximum output: 7 watts at 25 mph (40.23 kmph)
RPM at rated output: 430
Overspeed control: Non-destructive collapse in high winds
Generator/Alternator: 14 VDC, permanent magnet
Testing procedures: Moving test bed
Warranty: 30 days, parts and workmanship
Maintenance schedule: Monthly inspection